International Scientific and Practical Conference "Materials Science, Mechanical Engineering and Energy: Problems and Prospects for Development" (MSEE'2019)

IOP Conference Series: Materials Science and Engineering
Volume 665

Barnaul, Russia
27 - 28 June 2019

Editors:

Alexander A. Sitnikov

ISBN: 978-1-5108-9853-0
ISSN: 1757-8981

Printed from e-media with permission by:

Curran Associates, Inc.
57 Morehouse Lane
Red Hook, NY 12571

Some format issues inherent in the e-media version may also appear in this print version.

Institute of Physics
Dirac House, Temple Back
Bristol BS1 6BE UK

Phone: 44 1 17 929 7481
Fax: 44 1 17 920 0979

techtracking@iop.org

Additional copies of this publication are available from:

Curran Associates, Inc.
57 Morehouse Lane
Red Hook, NY 12571 USA
Phone: 845-758-0400
Fax: 845-758-2633
Email: curran@proceedings.com
Web: www.proceedings.com

Table of contents

Volume 665

International Scientific and Practical Conference "Materials Science, Mechanical Engineering and Energy: Problems and Prospects for Development"

27–28 June 2019, Polzunov Altai State Technical University, Russian Federation

Accepted papers received: 9 October 2019
Published online: 29 October 2019

Preface

Preface

EDITOR

Peer review statement

Papers

Methods for forming interfacial heat-mass exchange surface in flowing air contact with water

A.G. Averkin, A.I. Eremkin and S.O. Kiselev.....1

Studying microstructure of sintered magnets Co-25% Sm

O A Chikova, K Yu Shmakova and I V Slinkin.....6

Adhesion and mechanical properties of polyanionic cellulose-Al-Zr protective coatings

N M Antonova.....11

Realisation and Approbation of Conditionally Constant Coefficients Method for Loss Counter Measuring Tools

S.S. Kostinskiy, D.V. Shaykhutdinov, E.V. Kirievskiy, Y.M. Manatskov and N.D. Narakidze.....17

Peculiarities of the fracture upon tensile deformation of hypoeutectic silumin samples after irradiation with an intensive pulse electron beam digital image correlation method

Yu F Ivanov, A M Ustinov, A A Klopotov, D V Zagulyaev, E A Petrikova, Yu A. Abzaev, Yu A Vlasov and O M Loskutov.....23

Monitoring of properties of epoxy molding compounds used in electronics for protection and hermetic sealing of microcircuits

A N Kokatev, N M Iakovleva, K V Stepanova, N Yu Ershova and K V Belov.....30

Decision making system for industrial companies

A O Loyko and S A Gusev.....37

Aluminate spinel-based SHS pigments for fire-resistant protective and decorative coatings

N I Radishevskaya, A Yu Nazarova and N G Kasatsky.....42

Use of Additive Technologies to Optimize Design of Classifying Devices

A O Romashev.....48

Simulation and analysis of the shaft strength of a centrifugal pump in an in-situ leaching method of uranium

M P Sadykov.....54

Analytical description of stress-strain dependence for segmented polyurethane ureas swollen in physically aggressive media

V Yu Senichev, M A Makarova, E R Volkova, A V Savchuk and A I Slobodinyuk.....64

Magnetocaloric effect in thin magnetic films γ-Fe2O3

X O Urinov, X A Jumanov, Kh B Mirzokulov, A M Xidirov, M A Bolbekov, T A Jumaboyev and J O Urinov.....69

Martensitic transformations in VO2 films

X O Urinov, X A Jumanov, A K Amonov, A M Xidirov, J R Kilichov, J X Nurmurodov and T A Jumaboyev.....73

Workstations Industry 4.0 for instrument engineering products

D A Zakoldaev, A V Shukalov, I O Zharinov and O O Zharinov.....78

Workstations Industry 4.0 for instrument manufacturing

D A Zakoldaev, A G Korobeynikov, A V Shukalov and I O Zharinov.....82

The life cycle of technical documentation in the smart factory of Industry 4.0

D A Zakoldaev, A V Gurjanov, A V Shukalov and I O Zharinov.....87

IOP Conf. Series: Materials Science and Engineering **665** (2019) 011001 doi:10.1088/1757-899X/665/1/011001

Abstract

Conference «Materials Science, Engineering and Energy: Problems and Prospects of Development» (MSEE'2019)

PREFACE

The conference serves as a platform for discussing advanced and innovative technical solutions in various fields of modern materials science, engineering and energy. The purpose of the conference is the presentation of innovative approaches in the development of integration solutions in engineering and technology, as well as in interdisciplinary research. Key topics of the conference cover priority areas of research in the field of technical and related sciences and take into account global development trends.

MSEE'2019 topics are based on sustainability, with close attention paid to all forms of engineering professions and research tools, including the following topics:

1. New materials and technologies for their production

2. Design, processing and quality control of machines and mechanisms

3. Problems of traditional and alternative energy

4. Actual problems of modeling and automation of technological processes in mechanical engineering

5. Additive technology

6. Methods for creating coatings and thin films

IOP Conf. Series: Materials Science and Engineering **665** (2019) 011002 doi:10.1088/1757-899X/665/1/011002

EDITOR

Prof. Dr. Alexander A. Sitnikov

Polzunov Altai State technical University, Barnaul, Russian Federation

E-mail: msee_astu@mail.ru

Peer review statement

All papers published in this volume of *IOP Conference Series: Materials Science and Engineering* have been peer reviewed through processes administered by the proceedings Editors. Reviews were conducted by expert referees to the professional and scientific standards expected of a proceedings journal published by IOP Publishing.

IOP Conf. Series: Materials Science and Engineering **665** (2019) 012001 doi:10.1088/1757-899X/665/1/012001

Methods for forming interfacial heat-mass exchange surface in flowing air contact with water

A.G. Averkin[1], A.I. Eremkin[1] and S.O. Kiselev[1]

[1]Penza state University of architecture and construction,
28, Germana Titova str., Penza, 440028, Russia

e-mail: algraw@mail.ru

Abstract. A review of methods for forming interfacial heat-mass exchange surface in contact devices for air steam treatment is given. An effective method for creation of heat-mass exchange surface in air flow contact with water in the course of adiabatic wetting in air conditioning systems is proposed: by use of a plate head-piece with a mechanical vibration exciter. A method of defining area of the interfacial heat-mass exchange surface for these conditions is developed on the base of number of transfer units. The criterion equations to describe heat and mass transfer which occur when a plate head-piece with a vibration exciter is used for adiabatic air wetting are obtained. Use of a plate head-piece with vibration exciter as a contact device reduces energy consumption due to reduction of aerodynamic resistance of the air path and process intensification.

1. Introduction

In air conditioning systems, the following methods are used to form interfacial heat-mass exchange surface in the contact devices for steam treatment of air [3]:

1) spraying water in the air chamber using mechanical nozzles, for example, in the spraying nozzle chamber (SNC); in this case, interfacial heat-mass exchange surface area is equal to the total water droplets surface area.

2) air flow film contact with water film in the air chamber equipped with a head-piece (honeycomb wetting chamber). Under these conditions, the interfacial heat-mass exchange surface area is equal to the total surface area of all water films flowing down along the nozzle.

3) spraying water in the air chamber with mechanical nozzles and, in addition, crushing formed water droplets at contact with the plane of horizontal grids (spraying grid chamber - SGC) [2]. At that, the interfacial heat-mass exchange surface is equal to the total surface of water droplets formed due to nozzle spraying and water droplets formed additionally at the droplets crushing on the plane of the horizontal grid.

2. Body

According to Kokorin O. Ya. [3, p. 46], consumption of power for water spraying in the nozzle chamber is 25 times greater than in case of supplying water in the chamber with the head-piece. Therefore, method 2 of forming interfacial heat-mass exchange surface in contact devices for air steam treatment is the most effective, and honeycomb spraying chambers which use this method are increasingly used as contact devices in centralized air conditioners.

Layout of honeycomb spraying chamber is shown in Figure 1.

IOP Conf. Series: Materials Science and Engineering **665** (2019) 012001 doi:10.1088/1757-899X/665/1/012001

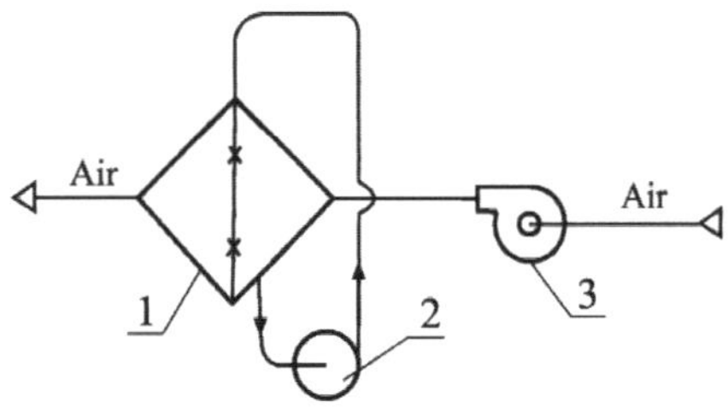

a

b

Figure 1. Layout of honeycomb spraying chamber:a – schematic diagram; b – general view of the honeycomb head-piece

The chamber is shaped as a rectangular parallelepiped and is equipped with a fan 6 for air supply. Throughout the height of the chamber assembly, a honeycomb head-piece made of corrugated hygroscopic plates 2 is installed.

The spraying water dispenser 3 made in the form of a perforated plate with through holes for uniform water distribution over the chamber (head-piece) section is mounted on the head-piece top. The lower part of the head-piece is located above the pallet of the chamber 4 which is used as a water tank. The chamber is equipped with a pump 5 which provides water recirculation: water from the pallet through the hose is fed to the spraying water dispenser; from the dispenser water enters the corrugated hygroscopic plates of the honeycomb head-piece. Water wets the plates of the head-piece and in the form of water film flows down by gravity into the pallet of the chamber. At that, the water temperature is stabilized at the level of the wet air thermometer temperature. Due to operation of the fan, air passes through the head-piece gaps, contacting the flowing water film, where it is adiabatically wetted (cooled), and leaves the chamber. The heat-mass exchange surface area between air and wetted surface is defined by the size (area) of the water film in the head-piece layer.

As it has been noted above, an advantage of adiabatic wetting (cooling) of air with water in the spraying layers is a significant reduction in power consumption by the water pump drive compared to the energy consumption in the nozzle chambers. In the honeycomb wetting chamber assembly, required pump pressure depends on the height of spraying water dispenser over the head-piece wetting assembly and is less by several times than the required pump pressure in the case when water is sprayed through mechanical nozzles. Also, decrease in power consumption for the water pump drive is due to decrease in the spraying coefficient – μ, i.e. water consumption per unit of treated air. [3].

Disadvantages of adiabatic air wetting with water in the honeycomb wetting chamber assembly include large air path aerodynamic resistance which increases energy consumption to drive the fan unit, and also increases the fan size.

To increase heat-mass exchange surface area between air and water in a unit volume of the honeycomb wetting chamber while reducing the aerodynamic resistance of the air path and, accordingly, improving efficiency of adiabatic air wetting and cooling at reduced specific water consumption for air treatment (spraying coefficient), the authors propose to use a plate vibration head-piece.

It is known [1, p. 212] that vibration effect on the working units, in particular, effect of targeted mechanical vibrations with a relatively small amplitude and frequency in the range 10...103 Hz, allows significant increasing the phases contact surface area and reducing diffusion resistance in mass transfer processes.

To achieve this technical result in the chamber assembly, the honeycomb head-piece was replaced by a plate head-piece assembly fixed to a spring vibration base and containing a mechanical vibration exciter with electric drive.

According to the mass transfer equation [3], increase in the heat-mass exchange surface area leads to the process intensification, i.e. to increase in the mass of steam passing into the air:

$$dM = k \cdot dF \cdot \Delta_{av} \cdot d\tau , \qquad (1)$$

where dM – increment of steam amount, kg;

k – mass transfer coefficient, s/m;

dF – increment of heat and mass transfer surface area, m^2;

Δ_{av} – average driving force of mass transfer, Pa;

$d\tau$ – time increment, c.

General view of the plate head-piece assembly with vibration exciter is shown in Figure 2.

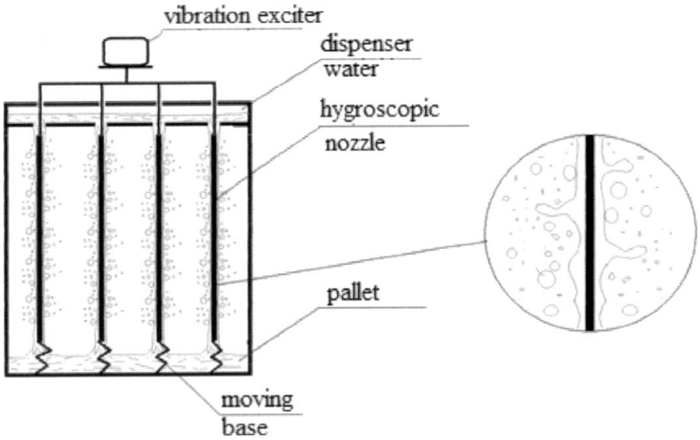

Figure 2. Plate head-piece with vibration exciter

The plate head-piece assembly consists of a set of flat parallel plates made of hygroscopic (or non-hygroscopic) material spaced at a distance of 15...20 mm and arranged along the length of the chamber.

The heat and mass transfer surface area between air and water is defined by a sum of:

a) surface area of the water film flowing down along the plate of vibration head-piece;

b) surface of the droplets and splashes in halo formed at water particles separation from the plates of the vibration head-piece.

It follows from analysis of Figure 2 that total area of heat and mass transfer increases significantly when a plate of vibration head-piece with a mechanical vibration exciter is used.

Aerodynamic resistance of the chamber assembly with a plate head-piece will be significantly reduced in comparison with the resistance in the honeycomb head-piece which is manufactured by foreign companies. Energy spent for air passage through the contact device is significantly reduced. In these conditions, additional energy consumption to drive the mechanical vibration exciter of the plate head-piece is small and can be of 200...300 W [1].

3. Methodical part

Calculation of heat and mass transfer surface area in the contact device which applies plate vibration head-piece is proposed using the method of graphic integration based on number of transfer units [4]:

1. Basic data for adiabatic wetting (cooling) of air (t_1, I_1, t_2 are, respectively, temperature, °C, enthalpy of inlet air, kJ/kg, and air temperature at the outlet of the contact device) were given.

2. The curve 1 – 2 for steam air treatment process was drawn in the *I-d* coordinate system, in accordance with Figure 3.

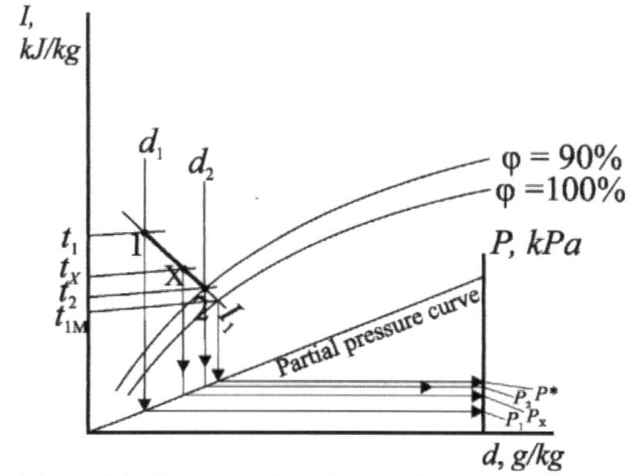

Figure 3. Defining current (t_i) and balance (t^*) values of water temperature, current air moisture content (d_i) and balance (P^*) values of partial steam pressures in the course of adiabatic wetting and air cooling.

3. Number of transfer units n_t necessary for exchange processes and number of transfer units n_p necessary for mass transfer processes were defined.

On the basis of Figure 3, a curve for adiabatic wetting (cooling) of air was drawn (Figure 4).

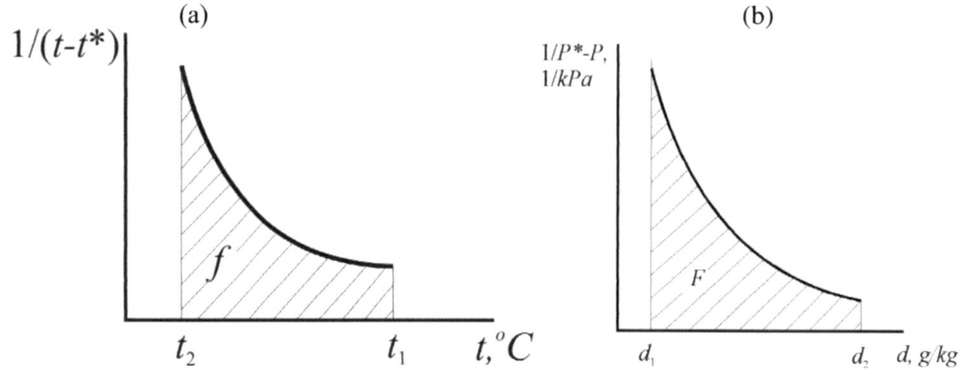

Figure 4. Curves for defining number of transfer units

Calculation of number of the transfer units can be made according to the formulas:

$$n_t = \int_{t_1}^{t_2} \frac{dt}{t-t_*} = f \cdot m_1 \cdot m_2 \quad (2)$$

$$n_p = \int_{d_1}^{d_2} \frac{d(d)}{P^*-P} = F \cdot M_1 \cdot M_2\,; \quad (3)$$

where t, t^* – respectively, current (operating) and balance values of air temperature during steam treatment, °C;

f, F – area of curved trapezoid in units of the used scale (Figure 4);

m_1, m_2, M_1, M_2 – respectively, scale of values along the abscissa and ordinate axes.

P, P^* – respectively, current (working) and balance partial pressure in the course of air wetting, kPa,

d_1, d_2 – current air moisture content, g/ kg.

4. Kinetic coefficients of heat and mass transfer were defined on the basis of thermal and diffusion similarity criteria.

Heat transfer coefficient is calculated by the formula:

$$\alpha = \mathrm{Nu}\frac{\lambda}{l}\,; \quad (4)$$

Mass transfer coefficient is calculated by the formula:

$$\beta = \text{Nu}`\frac{D}{l};\tag{5}$$

where λ – coefficient of thermal conductivity of air, W/(m·°C);

l – defining geometric dimension, m;

D - steam diffusion coefficient in air, m²/s.

Nu, Nu` – Nusselt criterion parameter for thermal process, diffusion process, accordingly, which are defined by a criterion equations obtained by the authors on the basis of experimental studies using multiple regression method.

$$\text{Nu} = 22.4\text{Re}^{0.11}\text{Re}_W^{0.41};\tag{6}$$

$$\text{Nu}' = 0.1\text{Re}^{0.464}\text{Re}_W^{0.195};\tag{7}$$

where Re, Re_W - accordingly, Reynolds criterion parameters for air flow, criterion for aqueous phase (film) in vibration mode.

$$\text{Re} = \frac{vd_э\rho}{\mu};\ \text{Re}_W = \frac{v_W d_э^* \rho_W}{\mu_W};\tag{8), (9}$$

where v. v_w – accordingly, air velocity and velocity of the water film in vibration mode, m/s; $d_э$, $d_э$* - respectively, equivalent diameter of the channel for air and equivalent diameter of the channel for water film, m; ρ, ρ_w – respectively, air density and density of the water film, kg/m³; μ_w μ, – dynamic viscosity coefficients for air and water film, Pa·s.

5. Required heat-mass exchange surface area in the contact device necessary for wetting in accordance with the basic data (p .1) was defined as:

$$F^t = \frac{n_t \cdot G}{\alpha};\tag{10}$$

$$F^p = \frac{n_p \cdot G}{\beta};\tag{11}$$

4. Conclusion

Application of a plate head-piece assembly with a mechanical vibration exciter in contact devices for air steam treatment provides an effective way to form interfacial heat-mass exchange surface in contact of gas (air) flow with water flowing down along the plate vibration head-piece in the form of a water film, as well as with water droplets and splashes in halo formed at water particles separation from the surface, i. e. to provide energy saving at air wetting (cooling) in adiabatic conditions.

Advantages of the considered method for forming interfacial heat-mass exchange surface include:

- improved efficiency of air steam treatment in the course of adiabatic wetting and cooling due to increased heat-mass exchange surface area between working environments;
- reduction of specific water consumption for air treatment (spraying coefficient);
- reduction of aerodynamic resistance of the air path.

References

[1] Varsonofiev V D, Kolman-Ivanov E E 1985 Vibrating technique in the chemical industry (Moscow: Chemistry)

[2] Golubkov B N, Romanova T M, Gusev V A 1988, Designing and operation of air conditioning plant (Moscow: Energoatomizdat)

[3] Kokorin O Ya 2005 Domestic equipment for ventilation and air conditioning systems (Moscow: CJSC "Printing house "Extra Pechat'")

[4] Averkin A G 2011 Stream treatment of air in ventilation and air conditioning (Penza: PGUAS)

[5] Radju K S 2016 Mechanics of fluid, heat transfer and mass transfer: chemical technology (Hoboken, New Jersey: John Wiley & Sons, Inc.,)

IOP Conf. Series: Materials Science and Engineering **665** (2019) 012002 doi:10.1088/1757-899X/665/1/012002

Studying microstructure of sintered magnets Co-25% Sm

O A Chikova[1,2] , K Yu Shmakova[1] and I V Slinkin[1]

[1]Ural Federal University, Mira, 19, Yekaterinburg, 620002, Russia, 2Ural State Pedagogical University, Russia, Yekaterinburg

e mail: chik63@mail.ru

Abstract. Metallographic study of sintered rare-earth Co- 25% Sm magnets microstructure of KS-25 brand was carried out by scanning electron force microscopy (EDS-analysis). A dendritic structure with three phases was found: branches of dendrites are $SmCo_5$; interdendritic space is a mixture of two phases $SmCo_5$ and Sm_2Co_{17}. Crystals of 1–5 μm of Zr_5Co_3FeSm compound and inclusions of Sm_2O_3 samarium of a globular form of 2–10 μm in size were founded also in the microstructure. Study of the domain structure on surfaces perpendicular to magnetization axis by magnetic force microscopy (MFM) showed presence of strong magnetocrystalline anisotropy. Comparison of magnetic images with electron microscopic images of surface made it possible to conclude that $SmCo_5$ dendrites correspond to large domains ~ 30–50 μm in size and the interdendritic space consisting of a mixture of two phases $SmCo_5$ and Sm_2Co_{17} correspond to a domain structure in labyrinth form with a size of ~ 3-5 μm.

1. Introduction

Sintered magnets based on the $SmCo_5$ intermetallic compound are used widely in engineering due to the large coercive force up to 40 kOe and magnetic energy up to 24 MGE [1-3]. Improvement characteristics of these magnets are impossible without a scientific study of principles formation of a highly coercive state, disclosure of a relationship between characteristics of microstructure, phase composition and magnetic properties. Microstructure of sintered $SmCo_5$ magnets includes fragile intermetallic compounds that can be destroyed both during manufacturing process and during operation. The study of the microstructure of magnets is also aimed at improving their mechanical properties.

The microstructure formation of sintered Co-25% Sm magnets occurs at several technological stages of manufacturing: 1) obtaining an alloy of the necessary composition, 2) preparing a fine powder, 3) pressing the powder in a magnetic field for formation of a texture, 4) fritting of powder billet and 5) carrying out thermal treatments to increasing of coercive force. Formation of a highly coercive state in sintered Co-25% Sm magnets is due to a change in surface of grains of main phase of $SmCo_5$. There are dispersed Sm_2Co_{17} precipitates on the surface of phase $SmCo_5$ grains during heat treatment process (slow cooling or low-temperature tempering) in the cobalt enriched magnets. Formation of Sm_2Co_{17} phase precipitates coherently connected with main $SmCo_5$ phase leads to a smoothing of grains surface, a decreasing in number of defective places that are nucleation centres of reverse magnetization domens and it increase coercive force in thermoheating magnets. To obtain high hysteresis properties sintered Co-25% Sm magnets should be somewhat enriched with samarium compared to stoichiometric of $SmCo_5$ composition. An excess of Sm is needed in order to compensate for its loss in manufacture, sintering of powders as a result of evaporation, oxidation and binding of Sm with impurities. Samarium enrichment is also necessary for formation of a non-equilibrium microstructure that ensures growth of coercive force in the sintered magnets during their heat treatment [4-5]. Study of local crystal structure of high-coercive $SmCo_5$ alloys by EXAFS spectroscopy using synchrotron radiation discovered to formation of disordered defects such as stacking faults during enrichment of alloys with samarium [5].

Machining of magnets (EDM and laser cutting, grinding) gives a deterioration in magnetic properties and even leads to destruction. Local heating and saturation of surface layers with hydrogen gives a decrease in magnetic properties of magnets and their destruction also. Sintered Co-25% Sm permanent magnets contain significant amounts of samarium rare earth element. Concentration of samarium at grain boundaries (in boundary phases, which are largely responsible for preservation of highly coercive state of magnets) reaches 70-98 wt.%. High affinity of rare-earth elements to oxygen and hydrogen is known, it leads to formation of oxides and hydrides. Under adverse climatic conditions of operation (high humidity and air temperature, in salt fog, hydrogen-containing atmosphere) magnetic and mechanical properties of magnets can significantly decrease, up to destruction. Additional mechanical loads in real conditions aggravate all this factors [6-7].

Structurally sensitive properties of magnetic material are initial and maximum magnetic perm abilities, coercive force, residual magnetization (residual induction) loss from hysteresis. These properties are extremely dependent on microstructure and sample size. The greatest influence on these properties is exerted by atoms of dissolved element, dislocations, grain boundaries, presence of second phase and its disparity. It is possible to influence these properties in a wide range by changing phase state and structure of alloys. Study of connection of the microstructure, the magnetic structure and the mechanical properties of magnets is theoretical and practical interest. The domain structure of magnetic materials mainly determines their magnetic properties. This work is devoted to the study of microstructure by means of scanning electron microscopy, magnetic structure by means of magnetic force microscopy of Co-25% Sm magnets of KS-25 brand in order to establish a connection between them.

2. Materials and Methods

Samples for study were selected from ready-made sintered magnets Co-25% Sm brand KS-25 according to GOST 21559-76. Chemical composition of the studied samples was Sm - 25 wt.%; Fe - 18wt.%; Cu - 9wt.%; Zr - 3wt.%; rest was Co. Before study the sample surfaces were grinded and polished by diamond abrasive down to 1 μm and colloidal silica during 20 min at the final stage.

A metallographic study was carried out on a Carl Zeiss AURIGA CrossBeam scanning electron microscope using the function energy dispersive spectroscopy (EDS). Scanning electron microscope AURIGA CrossBeam (Carl Zeiss, Germany) was equipped with the EDS system IncaEnergy (Oxford Instruments, UK) with 350X-MAX X-ray detector with spatial resolution about 1 μm and spectral resolution of 125 eV at Kα line of Mn. The SEM was used for visualization of surface morphology and structural defects in secondary electron mode with the resolution down to 2 nm. The EDS was applied for revealing of elemental composition of defects and inclusions. The EDS data collection and processing were made by means of IncaEnergy software.

Study of the magnetic structure was performed using the Asylum MFP 3D SA scanning probe microscope. Resulting images are spatial distribution of the Z-component of magnetic field on sample surface. Measurements were carried out using a two-pass technique, where the “magnetic” images were separated from the relief images. It was necessary to minimize influence of relief on the image of the distribution of magnetic forces. On first pass, surface relief was determined by the discontinuous-contact method. In second pass, on each scan line, the cantilever was raised above surface for a specified distance dZ, and scanning was performed in accordance with previously obtained relief. Thus, on second pass, distance between scanned surface and fixed end of the cantilever was constant. As a result, the image of relief and the magnetic image are obtained simultaneously. To observe spatial periodic domain structure it is sufficient to know derivative of magnetic interaction force; magnitude of magnetic force

$\vec{F} = (\vec{m}\vec{\nabla})\vec{H}$, where m is effective magnetic moment of probe, H is the stray field of sample.

Registered force of magnetic interaction and magnitude of field during movement of probe over the domain will be almost constant. In process of passing the cantilever over the domain wall, a slightly smoothed phase jump and amplitude of resonant oscillations are observed, which corresponds to a change in force [8-10].

3. Results and discussions

IOP Conf. Series: Materials Science and Engineering **665** (2019) 012002 doi:10.1088/1757-899X/665/1/012002

Results of microstructure studying of sintered magnet Co-25weight.% Sm brand KS-25 by means of scanning electron microscopy, including EDS analysis are presented in Fig.1. Data of EDS-analysis confirmed presence in magnets of a mainly single-phase $SmCo_5$ structure.

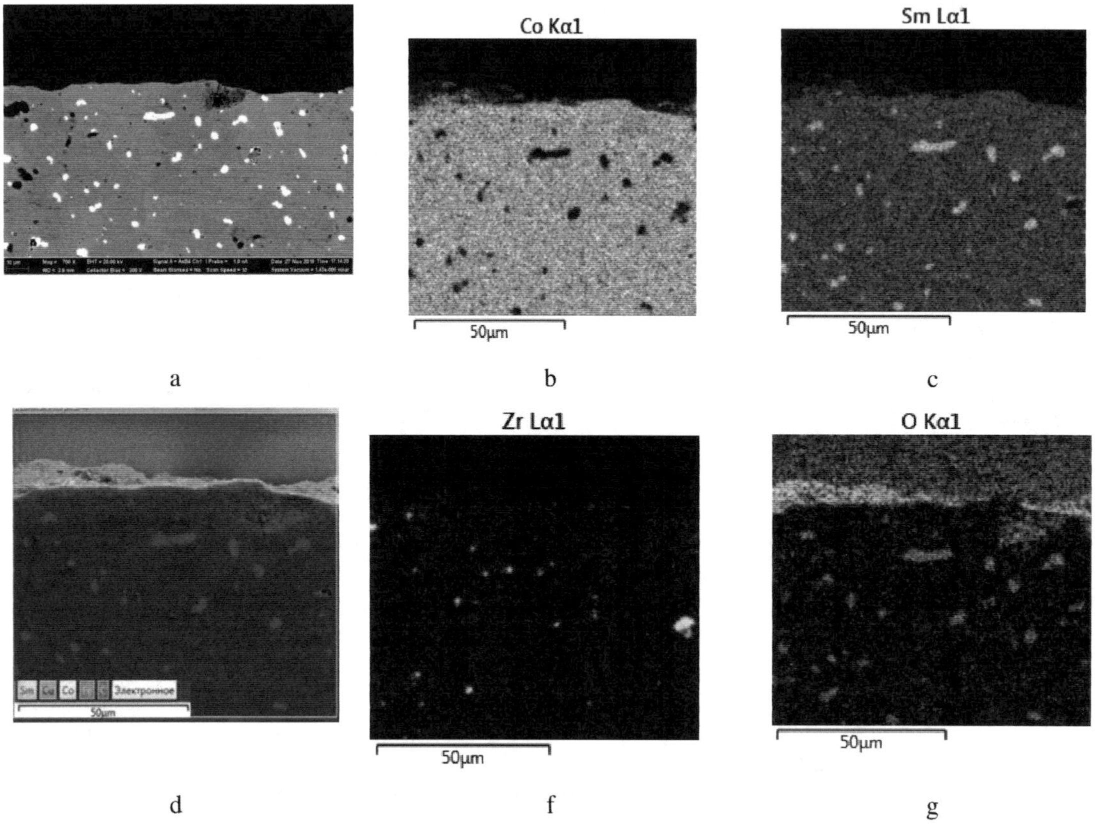

Figure 1. Microstructure of surface of sintered Co-25% Sm magnets of KS-25 brand. EDS-analysi in form of elements maps distribution (b c d e f g) and image (a) in the back scattered electrons

Microstructure is generally represented by three components: $SmCo_5$ dendrites (samarium content is 11at.%), the interdendritic space is a mixture of two phases $SmCo_5$ and Sm_2Co_{17} (samarium content 18 at.%), the crystals ~ 1-5 μm in size Zr_5Co_3FeSm (zirconium content is 47-52 at.%; cobalt is 25-28 at.%; iron is 11-13 at.%). Large inclusions of 2–5 μm samarium oxide Sm_2O_3 (oxygen content is 58–57 at.%, samarium content is 38 at.%) it is white globules in Fig. 1 were also found. The Sm_2Co_{17} phase has a lower hardness than the main phase of $SmCo_5$ (Sm_2Co_{17} microhardness is 600 MPa, $SmCo_5$ microhardness is 840 MPa) and is considered as a binding additive that improves mechanical properties. Sm_2O_3 inclusions are defects of metallurgical origin. Non-metallic inclusions of Sm_2O_3 sharply deteriorate mechanical properties of magnets; reduce strength and ductility, and corrosion sites appear in places of localization of these inclusions.

Results of the magnetic structure study on the surfaces perpendicular to axis of magnetization by magnetic force microscopy (MFM) are presented in Fig.2. Figure 2 displays spatial pattern of magnetic forces on surface sample. Images were obtained by measuring amplitude shift of the cantilever oscillation, which occurs under action of the magnetic force. The images contain information about location of the magnetic domains on sample surface. Grains of the main phase of $SmCo_5$ are dendrite branches with a characteristic size of ~ 20 μm in a magnetized state, they consist of one domain and correspond to lighter regions in Fig. 2. Observed domain structures are similar to those present in bulk uniaxial crystals with strong magnet anisotropy [11]. On surface perpendicular to axis of magnetization a domain structure is also observed in a labyrinth form. Small-scale domains have a transverse size of 3-5 microns and are magnetized perpendicular to surface. Their appearance is explained by a decrease in the magnetostatic energy due to higher energy of the domain wall. Large domain structures of grains, as a rule, are independent of neighbors but in some cases a significant magnetostatic connection between neighboring grains may be observed [12]. Connection between the domain width and the internal coercivity is also known. If the domain width decreases from 2 to 1 μm, energy of the domain wall also decreases, and the internal coercivity increases [13]. Comparison of magnetic structure images (Fig. 2) with electron microscopic images of surface (Fig. 1) made it possible to conclude that $SmCo_5$ (the dendrites branches are ~ 30-50 μm in size) consist of one domain, the mixture of two phases $SmCo_5$ and Sm2Co17 (the interdendritic space) have a domain structure in the labyrinth form with the size of magnetic phase components ~ 3 μm.

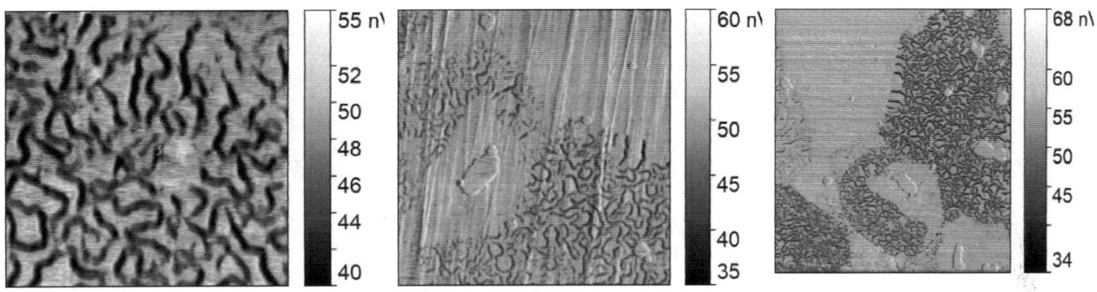

10x10μm 30x30μm 50x50 μm

Figure 2. MFM-image of surface of sintered magnets Co-25% Sm of KS-25 brand (amplitude of resonant oscillations)

4. Conclusions

Microstructure of sintered magnet Co-25% Sm of KS-25 brand is represented by SmCo5 dendrites (samarium content 11 at.%), the interdendritic spaces are a mixture of two phases $SmCo_5$ and Sm2Co17 (samarium content is 18 at.%). Microstructure also contains individual crystals of 1–5 μm of Zr_5Co_3FeSm compound (zirconium content is 47–52 at.%; cobalt is 25–28 at.%; iron is 11–13 at.%) and large globules 2–10 μm of samarium Sm_2O_3 oxide. Inclusions of Sm_2O_3 reduce strength and ductility and in localization places of inclusions corrosion pockets appear. Magnetic structure of sintered permanent magnets Co-25% Sm of KS-25 brand on surfaces perpendicular to axis of magnetization is represented by large $SmCo_5$ domains with a characteristic size of 30-50 microns and a domain structure in the labyrinth form from small-scale domains of 3-5 microns in size. Comparison of MFM-images with SEM-images of Co-25%Sm magnet surface made it possible to conclude that domains correspond to $SmCo_5$ dendrites, and domain structure in form of a labyrinth corresponds to an interdendritic space of a mixture of two phases $SmCo_5$ and Sm_2Co_{17}.

Acknowledgments

Authors are grateful for the support of experimental works by Act 211 Government Russian Federation, contract № 02.A03.21.0006.

References

[1] Aich S, Satapathy, D K and Shield J E 2017 Rapidly solidied rare-earth permanent magnets: Processing, properties, and applications *Advances in Magnetic Materials: Processing, Properties, and Performance* ed S Zhang, D Zhao (Boca Raton: CRC Press) chapter 7 pp 453-508

[2] Lileev A S and Menushenkov V P 1983 *Phys. Stat. Sol.* **5** 341
[3] Tarasov V P and Ignatov A S 2016 *Non Ferr. Met.* **2** 44
[4] Menushenkov V P, Sviridova T A et al 2009 *Proc. Int. Conf. on Metall. and Mater.* (Cerveny ZamekHradec Nad Moravici: Czech Republic/ Tanger Ltd) p.394
[5] Menushenkov A P, Menushenkov V P, Chernikov R V, Sviridova T A, Grishina O G, , Sidorov and K.V. Klementiev 2009 *J. Phys.: Conf. Ser.***190** 012091
[6] Lukin A A 1996 *Russ. Metall. (Metally)* **2** 131
[7] Nishio H, Yamamoto H, Nagakura M and Uehara M 1990 *IEEE Trans. on Magn.* **26** 257
[8] Mamin H J, Rugar D, Lambert S E, Franco L P, Sanders I L, Yogi T and Beaulieu T *J. Appl. Phys* **67** 5953
[9] Kazakova O, Puttock R, Barton C, Corte-León H, Jaafar M, Neu V and Asenjo A 2019 *J.Appl. Phys.* 125 060901
[10] Neu V, Vock S, Sturm T and Schultz L 2018 *Nanoscale* **10** 16881
[11] Szmaja W, Grobelny J, Cichomski M, Makita K and Rodewald W 2004 *Phys. Status Solidi* **201** 550
[12] Szmaja W 2007 *J. Magn. Magn. Mater.* **311** 496
[13] Wang Q, Wang Y, Li Y, Li P and Zhang X 2013*Chin. J. Rare Met.* **37** 896

Adhesion and mechanical properties of polyanionic cellulose-Al-Zr protective coatings

N M Antonova

Kamensk Institute (Affiliation) of SRSTU (NPI) named after M.I. Platov, 23 K. Marx Ave, Kamensk-Shakhtinsky, 347800, Russia

E-mail: melnik1@rambler.ru

Abstract. The paper reports on a new protective coating composed of polyanionic cellulose, Al and micro-additive Zr powders. The work has investigated a structure and phase composition of the composite coating. The study hasn't revealed any change of chemical composition in components of the overlay when coating and found macro-molecules of the polymer react with oxidic shells of particles due to hydrogen bonding. It has been established that adhesion strength of the coating increases 4 to 15 MPa when modifying an original suspension with zirconium micro-particles of less than 5 μm. The reason adhesion of the coating increases is reported to be agglomerates of Al and Zr fine fractions localized in hollows of the surface to be protected and on "adhesive-base" boundary.
Keywords: coating, adhesion, polyanionic cellulose, Al, Zr.

1. Introduction

To date, researchers have been focused on environmentally friendly materials with functional properties. Composites with a polymer matrix produced of renewable vegetable raw materials are considered to be prospective ones. The most applied of them are water dissolvable cellulose esters – carboxymethyl cellulose, methyl cellulose, ethyl cellulose, and polyanionic cellulose. Traditionally, these polymers are used as stabilizers, thickening and film forming agents [1-8]. Rheological properties of water dissolvable polymers have been studied in detail, so solutions with a certain viscosity and films with sufficient mechanical properties can be synthesized [9]. However, there is little published information on the development and application of protective coatings composed of biopolymers and dispersive metal fillers. Brittleness of biopolymer films and water dissolubility are major setbacks to the use of vegetable polymers.

In our previous research we demonstrated functional coatings could be synthesized with the use of carboxymethyl cellulose and dispersive metal fillers. The study has revealed that dispersion of filling particles and their distribution in the polymer matrix have a significant effect on structure formation in composite systems. It is possible to form composites with certain porosity, varying dimensions of particles to fill systems. The work [10] pointed at generation of orderly located pores in polymer films due to the modification of water-carboxymethyl cellulose-suspensions and Al powder (size < 45 μm) with highly dispersive particles AlOOH. These coatings demonstrated good anti-frictional properties in air when filling pores with solid greases [11]. No cell pores were found provided that the coatings were formed using carboxymethyl cellulose suspensions and Al without AlOOH [12]. Aluminum particles were distributed quite homogenously and compactly throughout the polymer matrix. The composites were used as corrosion protective coatings resistant to aggressive organic solvents.

Therefore, it is possible to produce materials with certain functional properties applying particles with different chemical characteristics and dispersion as filling agents. This study investigates into a new protective coating composed of polyanionic cellulose and aluminum powder. Rheological characteristics of polyanionic cellulose are more stable than those of carboxymethyl cellulose; moreover, this polymer is a good bonding agent for metallic particles. Zirconium powder with particles below 5 μm was used as a micro-additive component to increase adhesion strength of the coating. Previous research demonstrates that adhesion strength of coatings under consideration increases enormously due to added Zr, making them resistant to concentrated acids, alkalis and organic solvents. According to research data [13-15] Al and Zr improve mechanical and corrosion resistance properties of coatings. Nowadays, it is a burning issue to apply environmentally friendly coatings in order to replace non-ferrous metals through less expensive ferrous ones.

This work aims to get insight into the influence of zirconium powder on adhesion strength of an anticorrosion polyanionic acid – aluminum coating and reveal specifics of structure formation in the composite system under consideration.

2. Materials and Methods

Composite coatings containing purified polyanionic cellulose (a degree of polymerization 500), aluminum powder (produced through spraying of the melt) and a micro-addition of calciothermic zirconium were researched. As a plasticizer we used glycerin to be compatible with simple cellulose esters. SEM-data on powders used are given in Figure 1.

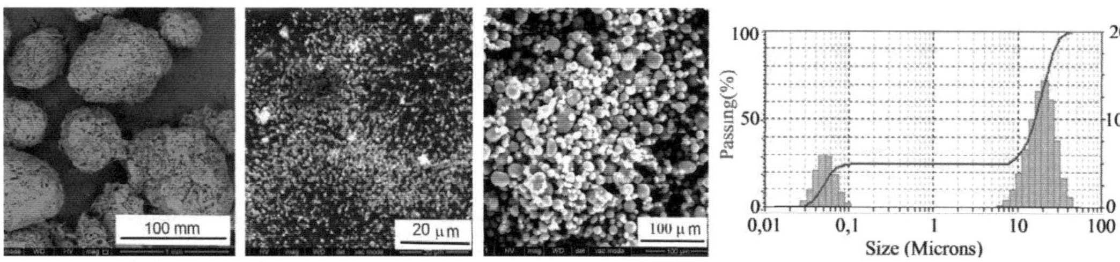

(a) (b) (c) (d)

Figure 1. SEM-data of powders (a) polyanionic acid, (b) Zr, (c) Al, (d) particle-size distribution histogram of Al powder.

The polymer has a fiber structure, in a dry powder fibers are like balls ranging to 900 μm in size. Zr powder contains roundish particles with a size of approximately 5 μm. Al particles are almost spherical ones with dimensions of about 45 μm. The histogram (Figure 1 d) shows data on dispersion of Al powder. A sub-micron fraction (≤0.1 μm) is found in aluminum powder. Peaks in the histogram are particles of approximately 19 μm in size. The biggest particles in the powder are in a range of 37÷44 μm, their percentage is 4.5 %.

2% gel-like water-solution of purified polyanionic cellulose powder, glycerin, and particles of Al and Zr powder were mixed to synthesize coatings. Distillated water was used. The coatings were deposited at a temperature of (40±1)°C. Sample coatings were composed with and without fillers (proportions of components were determined in previous research and given in Table 1).

Table 1. Proportions of components in suspensions of polyanionic cellulose water-solutions.

Compound	Polyanionic cellulose + glycerin + Al + Zr (g)
1	100.00+0+0+0
2	100.00+1.20+0+0

3	100.00+1.20+2.20+0
4	100.00+1.20+2.20+0.05

Structural morphology and element composition of coatings deposited on a steel base, physical configuration, dimensions of particles were researched using a SEM-microscope Quanta 200 in spectra of secondary and backscattered electrons. X-ray patterns of coatings were obtained using an X-ray diffraction meter RIGAKY ULTIMA IV in angle 2θ ranging 6° to 100° with an interval 0.02°. Voltage – 40 kV, anode current – 40 mA. Irradiation – CuK_{β}, K_{β}. Infrared spectra (IR - spectra) of organic components in composites and polymer powders were registered with the help of a Fourier spectrometer Varian 640 in a range 600÷4000 cm^{-1}. To measure dispersion of metallic powders we used an analyzer MicroTrac Bluewave. Research was carried out using facilities of Platov South-Russian State Polytechnic University. As the anticorrosion material under study is far less strong than a surface to be protected, we explored mechanical characteristics of coatings without the base material. Mechanical strength (σ) and relative strain were determined at axial tension (ε) of coatings deposited on a fluorpolymer base according to guidelines [16]. Adhesion strength of coatings on steel 08кп was analyzed using a gluing method [17]. The surface of samples was sandblasted for 5 minutes and degreased with acetone. When testing, a dynamometer PM-4 was used, so a speed of moving gripper along a fixed one was 2.5 mm/min.

3. Results and Discussion

The data on mechanical characteristics and adhesion strength of coatings are given in Table 2. Adhesion strength of polymer films without metallic fillers is low. Mechanical strength of compounds 3 - 4 is quite similar; however, the strength a coating bonds with the surface increases considerably (4 MPa to 15 MPa) provided that zirconium particles are added. Structure and phase composition of coatings were analyzed to find out a reason for higher adhesion strength of a surface to be protected with a coating when introducing zirconium particles. Figure 2 shows side surfaces of deposited coatings. Besides ordered zones some amorphous ones were detected in polyanionic cellulose film (Figure 2 a). There are also cracks; the film is brittle; so its adhesion and mechanical strength is low. An elastic film is formed via adding glycerin, but film-base adhesion drops (Figure 2 b). A coating with Al powder is amorphous, being quite homogenous indeed (Figure 2 c). A fine fraction of Al particles coagulates in a gel-ash polymer medium of the suspension, surrounding quite coarse particles of aluminum. As metallic micro-particles sediment and a solvent – water evaporates; a coating hardens with further tight packaging of aluminum particles in the polymer matrix. Packaging of particles in a composite containing powders Al and Zr is similar (Figure 2 d). Both coarse and fine Al particles are distributed quite homogenously on side surfaces in compounds 3 and 4.

Table 2. Mechanical properties and adhesion strength of coatings.

Compound	σ (MPa)	ε (%)	Adhesion strength (MPa)
1	9.0	10	1.0
2	11.2	15	0.8
3	18.0	12	4.0
4	19.2	10	15.0

(a) (b) (c) (d)

Figure 2. SEM – data of side surfaces of coatings: (a) polyanionic cellulose, (b) polyanionic cellulose + glycerin, (c) polyanionic cellulose + glycerin + Al, (d) polyanionic cellulose + glycerin + Al + Zr.

Figure 3 a shows IR–spectra of polyanionic cellulose powder and Al-Zr coating. Spectra are almost identical in a range 3400-3800 cm^{-1}. An absorption band in a range 3350-3200 cm^{-1} is in line with a valence fluctuating band of OH – group. This band for sample Al-Zr coating is more intensive than that for polyanionic cellulose sample. A band in range 1590 cm^{-1} identified in spectra corresponds with a group C=O for intermolecular hydrogen bonding. An absorption band tends towards lower frequencies because a hydroxyl group contributes to the formation of intermolecular hydrogen bonds [18]. This frequency for polyanionic cellulose sample is 3380 cm^{-1}, and for a coating 3275 cm^{-1}, respectively. The increase in intensity of absorption band for a group OH (3290 cm^{-1}) in a spectrum of the coating and dropping intensity of absorption band 1592 cm^{-1} for a group C=O as compared with a spectrum of powder polyanionic cellulose point at hydrogen bonding between active hydroxyl macromolecule groups and Al and Zr oxidic shells. A set of bands in a range 1350-700 cm^{-1} is an individual parameter for each substance [19]. They are nearly the same for polyanionic cellulose and a composite coating. The study on spectra revealed no transformation in a polymer matrix; chemical bonds are not formed between components of a coating.

X-ray pattern of Al-Zr coating (Figure 3 b) demonstrates diffraction peaks corresponding with peaks of a face-centered Al cubic cell (space group *Fm*3*m*) and reflexes similar to a six-sided tightly packed Zr cell (space group P63/mmc). A phase composition of original metallic fillers is constant when coating. Al and Zr oxides are not registered, probably, because of their low concentration in the coating.

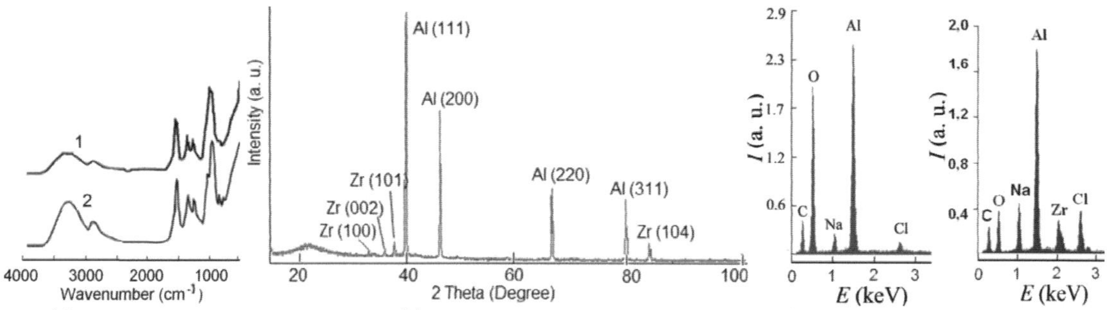

(a) (b) (c) (d)

Figure 3. (a) IR-spectra of polyanionic cellulose powder (1) and a composite polyanionic cellulose+Al+Zr+glycerin (2),
(b) X-ray pattern of a composite polyanionic cellulose+Al+Zr+glycerin; element composition of the surface in a composite (polyanionic cellulose + glycerin + Al + Zr): (c) formed in air, (d) deposited on the base.

Infrared spectroscopy and X-ray phase analysis haven't revealed any changes in phase composition of original components in the process of coating deposition.

The distribution of metallic particles on the contact edge "base – coating", and the way they fill up defects in the surface have a strong effect on adhesion strength of the composite. There are coarse particles in Al powder and also particles with a size similar to that of Zr particles (< 5 μm). Aluminum and zirconium particles have quite a similar spherical physical configuration. It is difficult to identify these particles using SEM methods. Therefore, we explored element composition in air-deposited Al and Zr compounds and that of a steel surface (Figure 3 c, d). According to the data of electron spectroscopy there is potassium, chlorine in the coating, so formation of NaCl is possible. This is not in contradiction with a traditional approach to the synthesis of highly purified polyanionic cellulose. NaCl is a secondary product in etherification reaction, when polyanionic cellulose is synthesized [18]. Producing highly-purified polyanionic cellulose NaCl is removed from the main substance by a solvent. A purified powder of polyanionic cellulose used in the study contains around 0.5% NaCl, as required in technical standards of the powder. The concentration of NaCl in the coating is low, so an X-ray pattern demonstrates no NaCl reflexes. A surface to contact with air contains Al (Figure 3, c). Al and Zr were detected on the surface close to the base (Figure 3, d). Therefore, Zr particles, a density of which exceeds that of Al, localize largely on the contact «base – coating» when they sediment. A peak-to-hollow distance in the surface under study (steel 08кп) ranges to 10 μm, so macro-roughness is registered. As Zr particles are mainly on the contact surface, adhesion strength of a coating with Zr increases, since hollows in a metallic base are filled up better with Zr particles. A coating-steel boundary consists of agglomerates formed due to sticking of Al and Zr particles in the coating and space. Agglomerates of fine Al and Zr fractions in the surface hollows and on the edge "adhesive – base" are like spanning elements, so adhesion increases due to mechanical bonding.

Resistance of compounds 3 and 4 was also checked, they are resistant to organic solvents, polyester varnishes, and quite stable to concentrated acids and alkalis.

4. Conclusions

To sum up, a new protective coating composed of polyanionic cellulose, Al and Zr powders is developed. A structure and phase composition of the composite coating has been investigated using methods of SEM and X-ray phase analysis and infrared spectroscopy. The study hasn't revealed any change of chemical composition in components of the overlay when coating. Macro-molecules of the polymer react with oxidic shells of particles due to hydrogen bonding. It has been shown that aluminum particles are packed tightly in a polymer matrix when hardening. Agglomerates of fine aluminum and zirconium fractions in the surface hollows and on the edge "adhesive – base" are like spanning elements, improving adhesion due to mechanical bonding.

References

[1] Alie A. O., Guo J., Wang S., Zhao X. 2016 *Journal of Natural Gas Science and Engineering.* **32** 491-500
[2] Yamaguchi K., Prabakaran M., Ke M., Gang X., Chung I.M., Gopiraman M., Kim I. S. 2016 *Materrials Letters.* **168**(1) 56-61
[3] Kierzek K. 2016 *Journal of Materials Engineering and Performance.* **25**(6) 2326-2330
[4] Marcia R. De Moura, Fause A. Aouada, Roberto J. Avena-Bustillos, Tara H. McHugh, John M. Krochta, Luiz H.C. Mattoso 2009 *Journal of Food Engineering.* **92** (4) 448-453
[5] Sahar Kafashi, Mohamadresa Rasaei. 2017 *Egyptian Journal of Petroleym.* **26** 371-374
[6] Nguen Q. D., Oh E.-S., Chung K. H. 2019 *Polymer Testing.* **76** 245-253
[7] Park Y.-S., Oh R.-S., Lee S.-M. 2014 *Journal of Power Sources.* **248** 1191-1196
[8] Patra J., Rath P.C., Li C., Kao H.-M., Wang F. M., Li J., Chang J. K. 2018 *ChemSusChem.* **11**(22) 3923-3931
[9] Kirsanov E. A., Matvienko V. N. 2016 *Neo-Newton behavior of structured systems* (Moscow: Technosphere)
[10] Antonova N. M., Babichev A. P., Berezovsky V. S. 2017 *Journal of Surface Investigation: X-ray, Synchrotron and Neutron Techniques.* **11** (5), 955-959
[11] Antonova N. M., Lisnichenko I, Sibirka I. MATEC Web of Conferences. 2017. **129** 02008
[12] Antonova N. M. 2009 *Russian Journal of Non-Ferrous Metals.* **49**(5) 419-423
[13] Kameneva A. L., Karmanov V. V., Dombrovsky I. V. 2014 *Research Journal of*

Pharmaceutical, Biological and Chemical Sciences **5**(6) 762-771
[14] Kameneva A. L., Karmanov V. V. 2014 *Journal of Alloys and Compounds* **546** 20
[15] Kameneva A. L. 2015 *Research Journal of Pharmaceutical, Biological and Chemical Sciences* **6**(1) 1381-1391
[16] Tushinsky L.I., Plokhov A.V., Tokarev A.S., Sindev V.I. 2004 *Methods of materials studies: structure, properties and processes of inorganic coating.* (M.: Mir Publishing).
[17] GOST (Russian State Standard) 14236-81. 1981 *Methods of tension testing* (M: Publishing of Standards).
[18] Petropavlosky G. A. 1988 *Gydrophilic Partially Substituted Cellulose Eteri and their Modification.* (by Chemical Cross-Linking. L.: Science)
[19] Belami L. 1963 *Infrared Spectra of Compound Molecules.* (M.: Foreign Literature Publishing Office)

Realisation and Approbation of Conditionally Constant Coefficients Method for Loss Counter Measuring Tools

Kostinskiy S.S., Shaykhutdinov D.V., Kirievskiy E.V., Manatskov Y.M., Narakidze N.D.

Platov South-Russian State Technical University (NPI) 132 Prosveschenia Str., Novocherkassk, Rostov Region, 346428, Russia

E-mail: d.v.shaykhutdinov@gmail.com

Abstract. To date, electrical power in a form of higher-order current harmonics and voltage are reported to be worsening substantially in power lines of industrial and non-commercial consumers. This fact relates to the change of power consumer types to use different household and industrial power receivers. Higher-order harmonic current components cause extra losses in transformers. As a consequence, electrical power delivery is less efficient and losses increase. Therefore, it is necessary to measure harmonic components of current and voltage in distribution networks. The paper presents a measuring tool of loss counter developed on the base of constant coefficients method. Performance and relative accuracy were studied experimentally using a variable-speed drive with the results reported in the paper. The developed measuring tool meets the standards of admitted relative accuracy and is applicable to implement conditionally constant coefficients method in order to calculate excessive losses to be caused by non-sinusoidal loads in transformers of distribution networks.

1. Introduction

Nowadays, electrical power consumption has been facing certain changes caused by a variety of domestic and industrial electrical receivers to use DC voltage generated via rectifying AC voltage and its leveling in a tank filter [1]. The equipment above creates non-lineal electrical loads for the power supply system and consumes non-sinusoidal electrical current. As a result, distortions are registered in supplying networks; furthermore, distorting voltage affects other machines to receive electrical power from one source [2]. It is said that electrical receivers are deteriorating electrical power much worse these days. Considerable quality degradation in power supplying lines of commercial and non-commercial consumers has been registered recently [3, 4]. The quality of electrical power in distribution networks is decaying in the most developed economies, since a smaller number of electrical receivers with linear volt-ampere parameters are left [5].

The supply of power is impossible without transformers, which transport it efficiently over distances and distribute among consumers. Being an important part in the energy system, transformers are found throughout all levels of voltage and considered to relate a network with consumers [6]. Higher-order harmonic components of electrical current cause extra losses in windings of transformers and excessive losses on vortex currents in a magnetic circuit. This fact has a negative effect on efficiency of power supply and cuts an estimated service life of electrical equipment and electrical networks due to accelerated heat and electrical aging of insulation [7]. Therefore, the rational use of power and reduction of losses in distribution networks with non-sinusoidal loads are challenging research and engineering issues in power engineering today [8].

Nowadays, operation modes of electrical power systems and networks appear to be characterized as special modes with non-sinusoidal and asymmetric loads, the introduction of which is dominated by new manufacturing technologies and domestic needs, so it is necessary to investigate how the non-sinusoidal load influences efficient power losses in power three-phase double-wound transformers [9, 10].

The study [11] researches the role of non-sinusoidal load parameters for losses of efficient power in power three-phase double-wound transformers using methods of constant coefficients. This paper suggests an algorithm to estimate losses of full transformer capacity applying the method of constant coefficients instead of a conventional method involving circuit equivalents of each harmonic. The algorithm above requires data on harmonic components of current and voltage in a network of 0.4 kV supplying power to consumers from transformers in a transformer station of a distribution network.

Harmonic components of current and voltage in distribution networks are to be measured with the purpose to determine the quality of electrical power. Today, a number of ready-to-apply approaches are available to analyze the quality of electrical power, e.g. quality analyzers: *MI* 2892, *Fluke* 437, version II, *CA* 8335 etc. The quality analyzers of electrical power above possess a variety of measuring functions, but they are expensive.

To sum up, it is necessary to develop a loss counter measuring tool based on conditionally constant coefficients in order to estimate extra losses caused by non-sinusoidal loads in transformers of distribution networks. The measuring tool of a loss counter is proposed to be constructed of available components, possess all required measuring functions and an acceptable cost price.

2. Materials and Methods

Basic parameters to be measured and estimated by a loss counter measuring tool are in line with characteristics of available quality analyzers of electrical power and listed in Table 1.

Table 1. – Basic parameters to be measured

Parameter, measurement unit	Measuring range	Limit of acceptable relative error, δ (%)
True RMS of AC voltage in phases *L*1, *L*2, *L*3, V	100 – 400	± 0.5
True RMS of current force in phases *L*1, *L*2, *L*3, A	0 – 5	± 1.5
True RMS of current force in a neutral conductor circuit, A	0 – 100	± 1.5
Amplitude of harmonic components of AC voltage in phases *L*1, *L*2, *L*3, B	1st to 50th	± 3 to ± 15
Amplitude of current force harmonic components in phases *L*1, *L*2, *L*3, A	1st to 50th	± 3 to ± 15
Effective power measured in one phase, W	0 – 2000	± 3
Power coefficient	0 – 1	± 3.5

Figure 1 shows a functional diagram of a loss counter measuring tool.

Further we outline the operation cycle of the loss counter measuring tool according to the functional diagram (Fig. 1), taking as an example one channel of current and voltage measurement in phase *L*1 (other current and voltage measuring channels in phases *L*2 and *L*3 are identical). The voltage signal in phase *L*1 comes to the resistance voltage divider 1, where it is decreased to the required level and transported further to a galvanic insulation amplifier 1. Power is supplied to the amplifier from an insulated *DC/DC* transformer 1. A signal, proportional to voltage to be measured in phase *L*1 is transported to input 1 of a synchronic analogue-to-digital converter (ADC). A signal of current in phase *L*1 is converted into a range 0 to 5 A by a current transformer, being changed further into a voltage signal ranging 0 to 1 V by a current clamp 1. A signal proportional to current to be measured in phase *L*1 comes to input 2 of the same ADC. Furthermore, there is an additional current clamp 4 to measure current in a neutral conductor circuit in a range 0 to 100 A.

To start the measuring process, *Arduino DUE* synchronizes time in real time module according to *GPS* signals. Time synchronization is carried out every 1.5 min. This way, measurements are synchronized when using several measuring tools of loss counters.

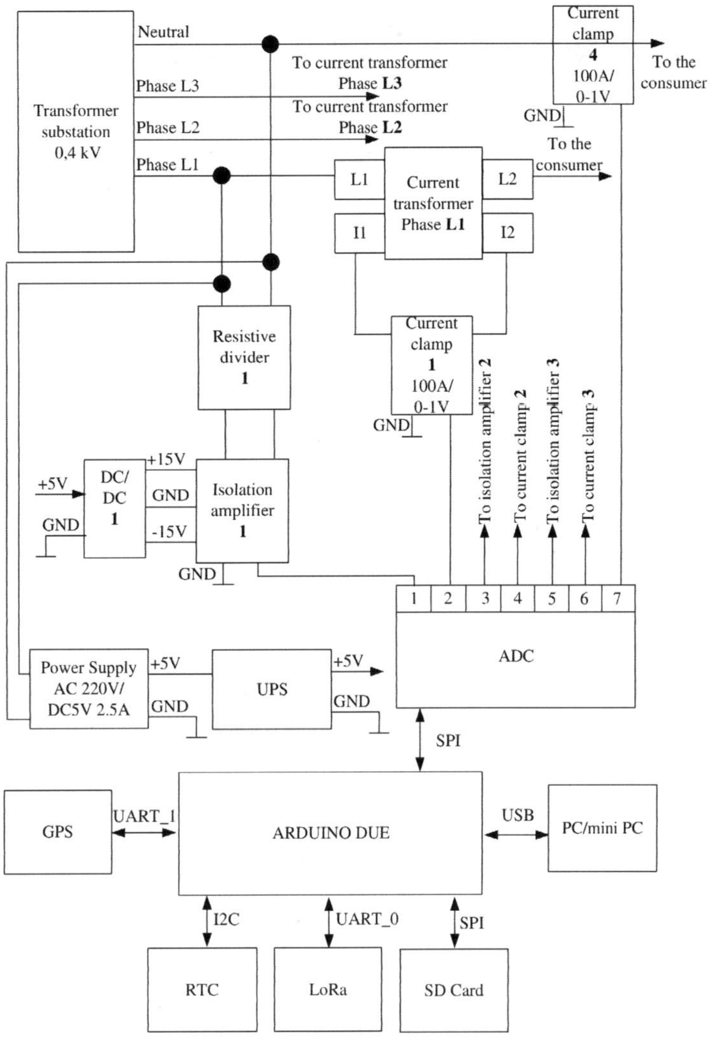

Fig. 1. – Functional diagram of a loss counter measuring tool

As module *Arduino DUE* gets the signal of interruption or even-not-even minute voltage numbers (proportional to voltage to be measured and current in phases *L*1, *L*2 and *L*3, as well as to current in a neutral conductor circuit) are recorded in a synchronal way according to interface *SPI* and ADC. Measuring lasts 0.2 s, i.e. 10 periods with a frequency of 50 Hz, each 0.02 s. The total number of points in one of seven ADC channels is 1000 (or 7000 in all channels) over the whole measuring period, or 100 points in 0.02 s (as stated in Kotelnikov theorem for a frequency of 50 Hz). *Arduino DUE* recalculates measured signals into instant values of current and voltage in the circuit according to ADC resolution, transformation coefficients of a current clamp and voltage divider. Further a true RMS of AC voltage current and current force in phases *L*1, *L*2, *L*3 is calculated, as well as a current force in neutral conductor circuit, efficient and full power, and a coefficient of power. Afterwards, discrete Fourier transformations are carried out to the 50th harmonic component of current and voltage [12]. All calculations and transformations done, the values are registered in a document «dd.mm.yyyy.csv» on *SD* card, providing time and date of record (time and date are taken as in a real time clock), and coordinates (longitude and latitude) according to *GPS* signals. Then the data similar to those saved on an *SD* card are transferred by *Arduino DUE,* using a *USB* bus to the PC or a pocket *Raspberry-based* PC. Simultaneously the data are *UART*-transferred by an interface to a wireless connection module *LoRa* for remote monitoring. After an even (or non-even minute) according to the real time clock, the cycle of measurement, calculation and transformation is repeated.

As module *Arduino DUE* gets the signal of interruption or even-not-even minute voltage numbers (proportional to voltage to be measured and current in phases *L*1, *L*2 and *L*3, as well as to current in a neutral conductor circuit) are recorded according to interface *SPI* and ADC. Measuring lasts 0.2 s, i.e. 10 periods with a frequency of 50 Hz, each 0.02 s. The total number of points in one of seven ADC channels is 1000 (or 7000 in all channels) over the whole measuring period, or 100 points in 0.02 s (as stated in Kotelnikov theorem for a frequency of 50 Hz). *Arduino DUE* recalculates measured signals into instant values of current and voltage in the circuit according to ADC resolution, transformation coefficients of a current clamp and voltage divider. Further a true RMS of AC voltage current and current force in phases *L*1, *L*2, *L*3 is calculated, as well as a current force in neutral conductor circuit, efficient and full power, and a coefficient of power. Further discrete Fourier transformations are carried out to the 50th harmonic component of current and voltage [12]. All calculations and transformations done, the values are registered in a document «dd.mm.yyyy.csv» on *SD* card, providing time and date of record (time and date are taken as in a real time clock), and coordinates (longitude and latitude) according to *GPS* signals. Then the data similar to those saved on an *SD* card are transferred by *Arduino DUE,* using a *USB* bus to the PC or a pocket PC on the base of *Raspberry*. Simultaneously the data are *UART*-transferred by an interface to a wireless connection module *LoRa* for remote monitoring. After an even (or non-even minute) according to the real time clock, the cycle of measurement, calculation and transformation is repeated.

The developed measuring tool of the loss counter is supplied from 220 V network through a power supply unit and uninterruptible power supply unit. A voltage of power is 5 V, current ranges to 2.5 A.

3. Results and Discussion

Experiments to determine performance and relative accuracy of the developed loss counter measuring tool were carried out with the help of a variable-speed drive *VLT* 6000 manufactured by *Danfoss*. The quality analyzer of electric power made by *METREL* (*MI*2792) was used as a sample device.

The following conclusions were made according to the results of measurements. A maximal absolute measurement error in phase *L*1 for voltage doesn't exceed 0.5 V, for current force – 0.0195 A, efficient power – 3 W, power coefficient – 0.02, therefore, relative accuracy is below 0.36 % for voltage; 1.36 % for current; 2.5 % for efficient power, and 3.33 % - power coefficient.

A maximal absolute measurement error in phase *L*2 for voltage doesn't exceed 1.1 V, for current force – 0.0215 A, efficient power – 8.4 W, power coefficient – 0.02, therefore, relative accuracy is below 0.79 %; for voltage; 1.49 % for current; 3.0 % for efficient power, and 3.33 % - power coefficient.

A maximal absolute measurement error in phase *L*3 for voltage doesn't exceed 0.3 V, for current force – 0.0202 A, efficient power – 3.4 W, power coefficient – 0.02, therefore, relative accuracy is below 0.22 % for voltage; 1.5 % for current; 2.96 % for efficient power, and 3.33 % - power coefficient.

As the level of even harmonic components was quite low when measuring, the results are omitted. The data on non-even measurements of current force harmonic components (n) 1 to 25 in phases L1, L2 and L3 are given in Figures 2, 3 and 4 (harmonic components after 25 is quite low and is not shown).

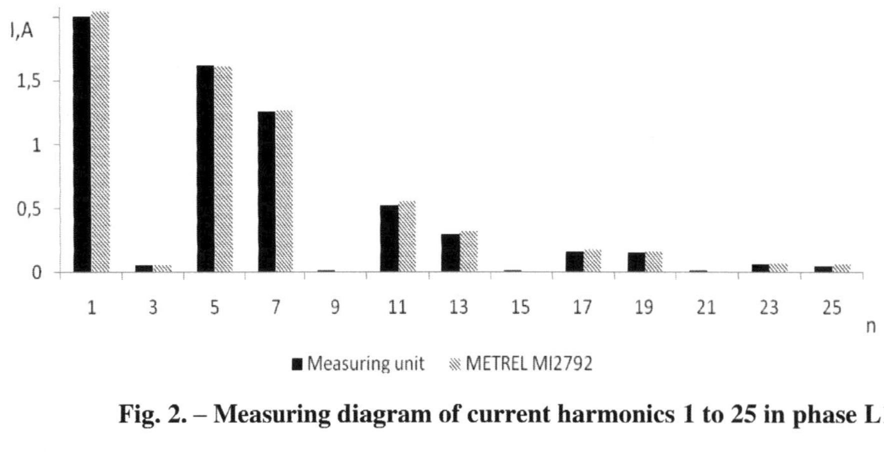

Fig. 2. – Measuring diagram of current harmonics 1 to 25 in phase L1

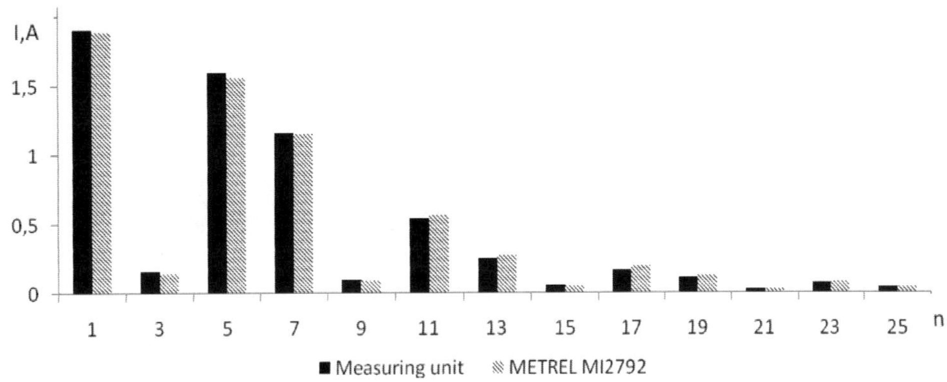

Fig. 3. – Measuring diagram of current harmonics 1 to 25 in phase L 2

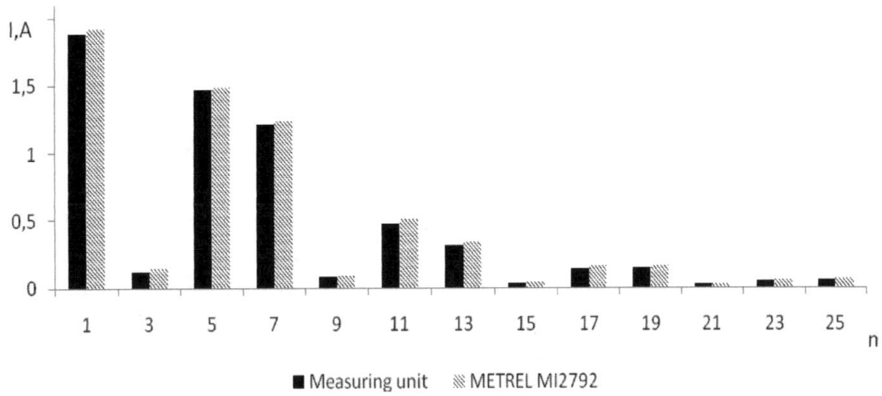

Fig. 4. – Measuring diagram of current harmonics 1 to 25 in phase L 3

As seen in the measurements above, maximal relative accuracy of measured harmonic components of current force in phase *L*1 is below 13.61 %, in phase L2 – 13.87 %, in phase *L*3 – 14.56 %.

4.Conclusion

The developed loss counter measuring tool determines parameters listed in Table 1 according to specified relative accuracy and can be used to implement the developed method of conditionally constant coefficients of excessive losses caused by non-sinusoidal loads in transformers of distribution networks.

Acknowledgments
The reported study was funded by RFBR according to the research project № 18-38-20188 «Development and research of tools for intellectual modeling and synthesis of electric power networks technological processes trajectories» using equipment of shared facility «Diagnosis and energy-efficient electrical equipment» (NPI).

5. References

1. Singh R. Aging of distribution transformers due to harmonics [Electronic resource] / R. Singh, A. Singh // Proceedings 14-th international conference on harmonics and quality of power (ICHQP). – 2010. – art. № 5625347.

2. Cazacu E. Derating the three-phase power distribution transformers under nonsinusoidal operating conditions: a case study [Electronic resource] / E. Cazacu, L. Petrescu // Proceedings of international conference on harmonics and quality of power (ICHQP). – 2014. – art. № 6842930. – P. 488 – 492.

3. Fuchs E. Power quality in power systems and electrical machines [Text] / E. Fuchs, M. Masoum Academic press. – 2008. – 638 p

4. Neha, C.K., Hegde, M., Aher, V., Effect of home appliances on power quality of conventional grid. In: 2016 International Conference on Circuits, Controls, Communications and Computing (I4C), art. № 8053260, pp. 1-6. IEEE (2016).

5. Cazacu E. Losses and temperature rise within power transformers subjected to distorted currents [Electronic resource] / E. Cazacu, L. Petrescu, V. Ionita. // Proceedings 15-th international conference on electrical machines, drives and power systems (ELMA). – 2017. – art. № 7955464. – P. 362 – 365.

6. Faiz J. Derating of transformers under non-linear load current and non-sinusoidal voltage – an overview [Text] / J. Faiz, M. Ghazizadeh, H. Oraee // IET Electric Power Applications. – 2015. – Vol. 9 (7). – P. 486 – 495.

7. Arsov L. Measurement of the influence of household power electronics on the power quality [Electronic resource] / L. Arsov, I. Iljazi, S. Mircevski, M. Cundeva-Blajer, A. Abazi // Proceedings 15-th international power electronics and motion control conference and exposition (EPE-PEMC). – 2012. – art. № 6397239. – P. DS1d.71 – DS1d.77.

8. Kostinskiy, S.S.: The functional dependence for the estimation of additional losses of active power in a double-wound power transformer caused by asymmetric active-inductive load with a delta connection. International Electronic Journal of Mathematics Education 11.6, 1529–1544 (2016).

9. Liu, C.W., Luo, C.C., Lin, P.Y.: Develop a power quality measurement system integrated with HAN home energy management system. In: 4th international conference on electric utility deregulation and restructuring and power technologies (DRPT), art. № 5994135, pp. 1506–1510. IEEE (2011).

10. Mog, G.E., Ribeiro, E.P.: Total harmonic distortion calculation by filtering for power quality monitoring. In: 2004 IEEE/PES Transmision and Distribution Conference and Exposition, IEEE Cat. No. 04EX956, pp. 629-632. IEEE, Latin America (2004).

11. Kostinskiy, S., Troitskiy, A.: Impact study of non-sinusoidal load parameters on real-power losses in 3-phase double-wound power transformers: method of constant coefficients. Journal of Engineering and Applied Sciences 12.20, 5058-5068 (2017).

12. Winograd, S.: On computing the discrete Fourier transform. Mathematics of computation 32.141, 175–199 (1978).

Peculiarities of the fracture upon tensile deformation of hypoeutectic silumin samples after irradiation with an intensive pulse electron beam digital image correlation method

Yu F Ivanov[1], A M Ustinov[1,2], A A Klopotov[2,3], D V Zagulyaev[4], E A Petrikova[1], Yu A. Abzaev[2], Yu A Vlasov[2], O M Loskutov[2]

[1]Institute of High-Current Electronics SB RAS, Tomsk, 2/3 Academichesky Ave, 634055, Russia
[2]Tomsk State University of Architecture and Building, 634003, Tomsk, 2 Solyanaya Sq, 634003, Russia
[3]National Research Tomsk State University, Tomsk, 36, Lenin Ave, 634055, Russia
[4]Siberian State Industrial University, Novokuznetsk, 42, Kirov Street, 654080, Russia

e-mail: klopotovaa@tsuab.ru

Abstract. The deformation behavior of non-irradiated and irradiated (with an intense pulsed electron beam) samples of hypoeutectic silumin ASTMB179-92A has been investigated. The distribution of localized plasticity foci in the near-surface layers of the samples has been studied. The dependence on the state of the material (non-irradiated and irradiated samples) and the distribution of deformation fields on the surface of silumin ASTMB179-92A under tension have been revealed. A correlation between deformation stages and the evolution pattern of spatial foci of the localized deformation has been established, and the staging of deformation curves has been revealed on the non-radiated samples. It has been shown that staging of the deformation process does not manifest itself on deformation curves of the irradiated samples. Deformation curves are characterized by a parabolic functional dependence $\sigma=f(\varepsilon)$. It has been established that irradiation with an intense pulsed electron beam leads to hardening of samples, and fracture is observed at higher values than on non-radiated samples. A significant deformation localization in the region corresponding to the nucleation of the crack appears at stresses close to brittle failure of samples under tension.

1. Introduction

Investigation of mechanical properties of aluminum-based alloys is initiated by intensive development of many branches of modern industry [1]. Recently, methods of the material surface modification with concentrated energy fluxes, including high-energy pulsed electron beams, have been actively used in the development of alloys with an increased complex of mechanical properties [2]. As a rule, modification of the surface layer of a material with concentrated energy fluxes is accompanied by formation of a gradient structure, the properties of which depend on the distance to the treatment surface. This circumstance significantly complicates the analysis of the deformation process of such materials. One of the directions for the analysis of the localization of plastic deformation of the material at meso- and macro-scale levels is the speckle interferometry [3], implemented in works [4, 5] with the use of a three-dimensional digital optical system Vic-3D.

The purpose of the article is to identify and analyze distribution patterns of local deformation foci under tensile conditions of flat samples from hypoeutectic silumin ASTMB179-92A in the initial, and the irradiated with an intense pulsed electron beam, states.

2. Materials and methods of the experiment

Samples of ASTMB179-92A silumin were the research material. Proportional flat samples with a tail for cantilever fastening were used in tensile tests. Test samples were divided into two batches. The first batch consisted of silumin samples in a cast state. The samples of the second batch were additionally irradiated with an intense pulsed electron beam using the SOLO installation [6] with the following parameters: the pulse duration of the electron beam is 150 μs, the number of pulses is 3, the pulse repetition rate is 0.3 s^{-1}, the electron beam energy density is 25 J/cm^2.

The tensile test was carried out using an INSTRON 3386 testing machine. The deformation rate was constant and was equal to 2 mm/min. The deformation distribution in the near-surface layers of the sample under tension was obtained using the VIC-3D optical measuring system [7, 8]. The image recording scheme on the surface of samples for obtaining digital stereoscopic images under tension is shown in figure 1. Before tests, speckle structures were created on the surface of samples with contrasting fine white and black matte aerosol paints. Changes in displacement fields during the deformation process in real time were obtained based on processing of digital stereoscopic images using the VIC-3D optical measuring system. These displacement fields represent displacement projections of local regions of the sample surface along the OX axis (transverse deformation) and along the OY axis (longitudinal deformation). During further processing, using the software of the VIC-3D system, the displacements were converted to relative deformations (ε_{xx} is along the X axis, ε_{yy} is along the Y axis, ε_{xy} is the shear deformation). The deformation pattern of the sample surface is obtained by combining changes in microregions. Structural studies of the surface of the ASTMB179-92A alloy were carried out using scanning electron microscopy (LEO EVO 50 (Zeiss, Germany)).

3. Research results and discussion

The structure of ASTMB179-92A silumin in the cast state is characterized by the presence of micropores, inclusions of intermetallides and silicon of micron sizes (figure 2). The presence of micropores and inclusions (especially plate-shaped) contributes to embrittlement of the alloy.

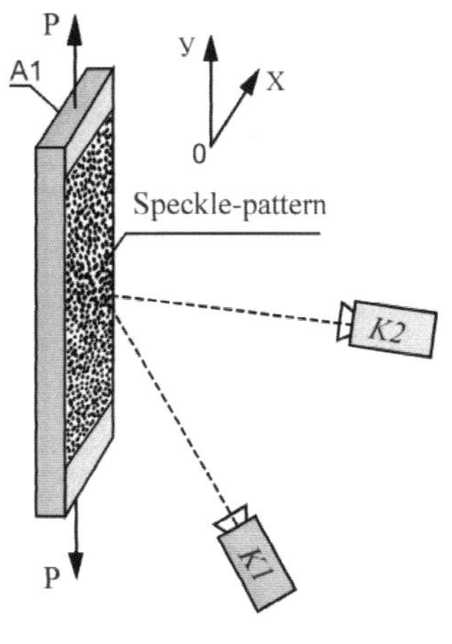

Figure 1. The scheme of image registration of the a speckle pattern on the lateral surface of the sample under tension; K1 and K2 are digital cameras; P is the applied load.

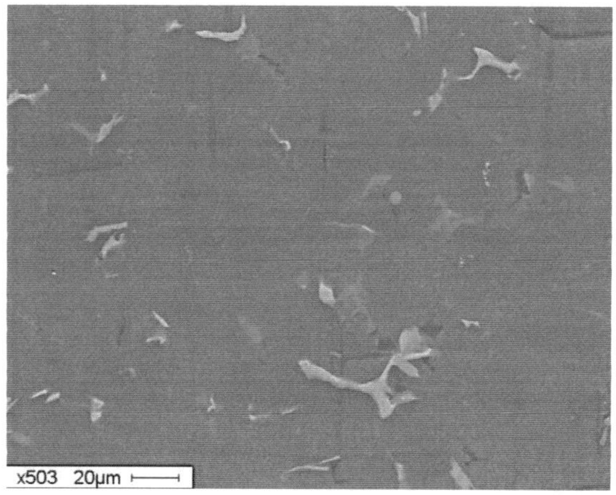

Figure 2. The surface structure of the hypoeutectic silumin ASTMB179-92A sample, identified using methods of scanning electron microscopy. The initial state.

First tensile stage (figure 3, *a*, stage I) is characterized by elastic deformation. At this stage, in distribution patterns of horizontal and vertical relative deformations, plastic deformation foci are localized in small regions of the sample surface (figure 4). Points 1 and 2 on the deformation curve $\sigma=f(\varepsilon)$ at stage I (figure 3) correspond to the distribution pattern of horizontal and vertical relative deformations on the silumin plate surface (figure 4). These regions are randomly located along the surface of the sample with the presense of both positive and negative values. A combined local deformation foci ε_{YY} with compression and tension regions (pattern 1, figure 4) has formed in the central part of the sample, which then, with increasing applied external stress, has transformed into a local deformation foci of only tension (pattern 4, figure 4).

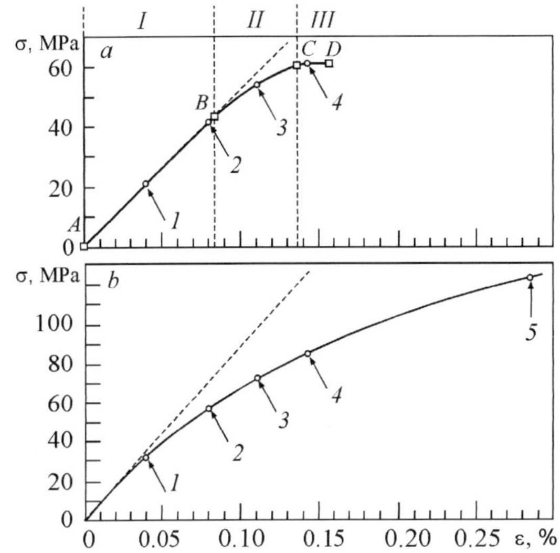

Figure 3. Deformation diagrams of non-irradiated (a) and irradiated (b) silumin with an intense pulsed electron beam. The numbers correspond to position of the deformation structure patterns on the curve (figure 4 and figure 5). AB is the stage of elastic deformation (stage I); BC is the transitional stage (stage II); CD is the stage of deformation softening (stage III) and fracture.

Distribution of deformations ε_{YY} in the vertical direction along the OY axis changes at stage II (figure 3, *a*). Individual plastic deformation foci merge in the lower and upper parts of the samples, (pattern 3, figure 4). Moreover, at stage II, plastic deformation foci (with $\varepsilon_{YY}<0$) almost completely disappear, and plastic deformation foci (with $\varepsilon_{YY}>0$) remain (figure 4).

The analysis of ε_{XX} deformation distribution patterns in the horizontal direction shows that the local deformation foci along the OX axis have an elongated shape, mainly in the direction along the vertical axis of the sample (the OY axis, figure 1). While foci of relative local deformation ε_{YY} have an elongated shape, mainly in the direction perpendicular to the vertical axis of the sample (axis OX, figure 1).

Transition from stage II to stage III ends with the fracture of the sample (figure 3, *a*). At stage III, in the central part of the sample, the deformation ε_{YY} in the local focus is more than twice the average deformation over the entire sample. Further destruction of the sample occurs in this area.

It has been shown that elastic deformation stage reaches values of ~0.09% under tension of non-irradiated samples of hypoeutectic silumin ASTMB179-92A. In the literature, the margin for residual deformation for steels and aluminum alloys is usually 0.2% (plastic deformation is 0.002) [9]. Thus, the obtained results characterize the investigated silumin as very fragile.

Further analysis of the obtained results has been carried out on the basis of constructing deformaton diagrams for maximum values of vertical and horizontal deformations inside plastic deformation foci on the surface of the samples (figure 5). The deformation diagram, with respect to maximum values f vertical deformations inside plastic deformation foci on the surface of the samples under tension, has a functional dependence $\sigma=f(\varepsilon_{YY})$, close to parabolic (with the exception of the section AB, figure 5, *a*). Plastic deformation of the material in these foci is more than 3 times greater than the average deformation of the sample at the last stage III.

The deformation diagram, with respect to maximum values of horizontal deformations inside plastic deformation foci on the surface of the samples under tension, is shown in figure 5, *b*. It is characteristic that the functional dependence $\sigma=f(\varepsilon_{XX})$ can be represented as two linear segments AC and CD.

The stress at point C in the diagram (figure 5, *b*) coincides with the stress in the deformation diagram of the alloy under study, at which the transition from stage II to stage III begins (point *C*, figure 3) and correlates with the stress at point C in the deformation diagram, relative to maximum values of vertical deformations (figure 5, *a*).

Thus, a difference in functional dependences $\sigma=f(\varepsilon_{YY})$ and $\sigma=f(\varepsilon_{XX})$ of deformations in plastic deformation foci has been revealed during tensile deformation of cast silumin samples. The dependence of the deformation behavior of the material on the deformation direction has been revealed.

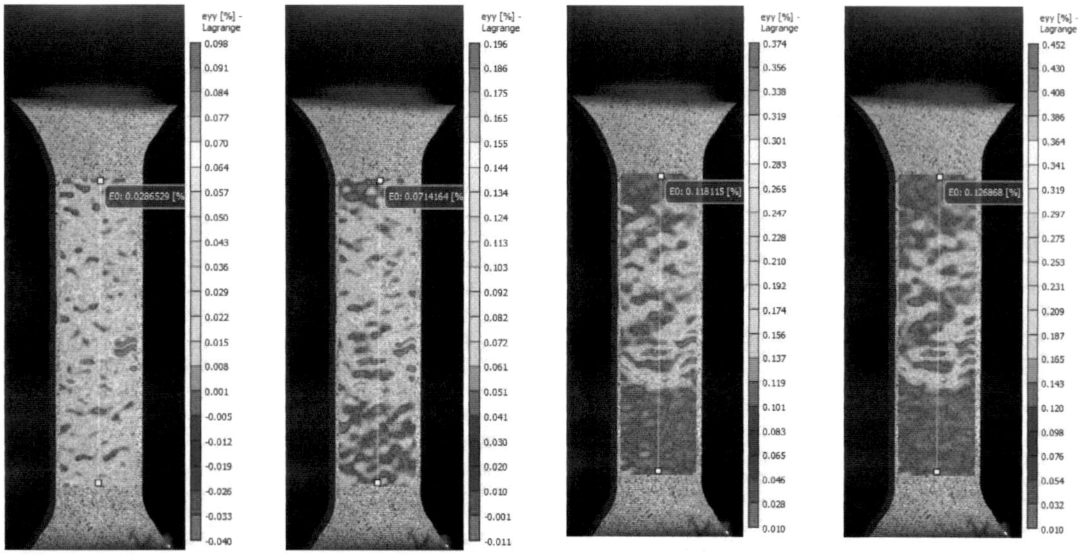

1 2 3 4

Figure. 4. Distribution patterns of vertical relative deformations on the surface of a flat plate of cast siluminin ASTMB179-92A. Figures 1, 2, 3, 4 in the diagram in figure 3 *a* show the corresponding strain-stress states of the sample during the test.

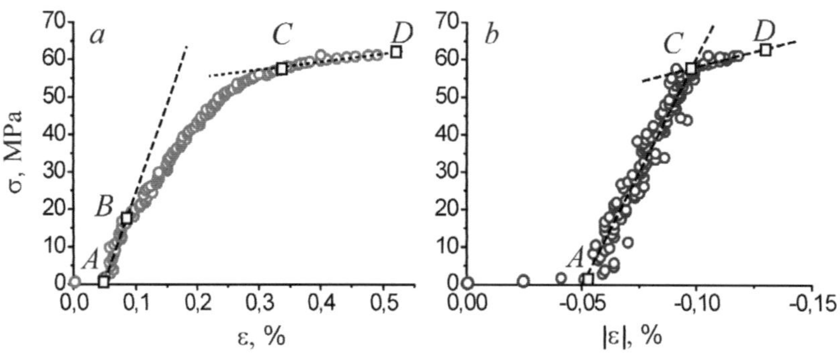

Figure 5. Deformation diagrams, with respect to maximum values of vertical $\sigma=f(\varepsilon_{YY})$ (*a*) and horizontal $\sigma=f(\varepsilon_{XX})$ (*b*) deformations inside plastic deformation foci on the surface of non-radiated silumin samples under tension. AB and AC are linear sections on the dependences σ - ε and σ - |ε|, respectively.

The above mentioned patterns (figure 4) of relative deformation distributions on the surface of a flat plate in the middle part of the sample clearly exhibit significant plastic deformation foci beginning from ~ 20 MPa. This phenomenon reflects the development of the plastic deformation localization in this particular part of the sample. With further tension, the appearance of plastic deformation foci in other parts of the sample suggests that stress concentrators exist in these places in the sample. However, an increase in the applied voltage leads to development of the local deformation region in the central part of the sample (figure 4). Ultimately, this leads to formation of cracks and brittle fracture of the sample as a whole.

The results of the tensile test of the irradiated sample of ASTMB179-92A silumin are shown in figure 3 *b*. The resulting diagram is characterized by a parabolic functional dependence σ=f(ε) [10]. It is impossible to single out deformation stages, as it had been done on the deformation curve of the non-irradiated sample (figure 3, *a*). Irradiation with an intense pulsed electron beam leads to sample hardening, and fracture is observed at higher values than on non-radiated samples. At stresses close to brittle failure of the samples under tension, in the formed localization foci of the plastic deformation, the value of the plastic deformation in horizontal direction (along the OX axis, figure 6, *d*) is close to total deformation of the entire sample. Whereas, in plastic deformation localization foci on the sample surface in the vertical direction (along the OY axis, figure 6, *c*), the magnitude of plastic deformation values is significantly less than the total deformation.

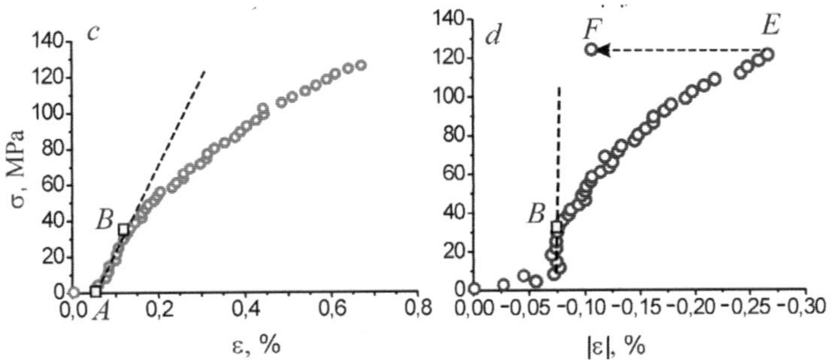

Figure 6. Deformation diagrams for maximum values of vertical $\sigma = f(\varepsilon_{YY})$ (*c*) and horizontal $\sigma = f(\varepsilon_{XX})$ (*d*) deformations inside plastic deformation foci on the surface of irradiated silumin samples under tension. AB and AC are linear sections on dependences σ from ε and σ from |ε|, respectively. EF is the line reflecting the stage of a sample fracture on the irradiated sample.

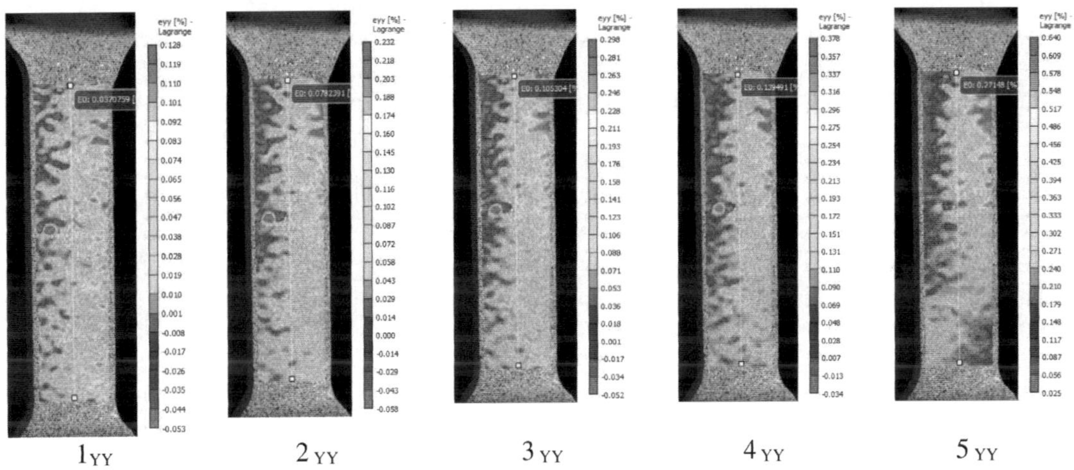

Figure 7. Distribution patterns of vertical relative deformations on the surface of a flat plate of the irradiated silumin ASTMB179-92A. Numbers 1, 2, 3, 4, 5 in the diagram show in figure 3, *b* the corresponding strain-stress states of the sample under test.

The staging of the plastic flow curve is associated with the nature of the plastic deformation localization [11]. It is impossible to distinguish individual stages for the non-irradiated alloy on the deformation curve, which is reflected in a similar form of deformation distribution patterns on the sample surface (figure. 3, *b* and figure 7).

It is obvious that such an evolution of deformation patterns of the deformation distribution on the surface of the irradiated sample in the process of tensile deformation reflects the presence of a hardened surface layer on the sample [12]. When studying deformation maps, it can be seen that the elastic deformation foci, which then transforms into plastic deformation foci, are localized at the initial moment on the lateral faces of the sample. Further, fusion of these plastic deformation foci is observed with an increase in the applied stress. As a result, the deformation field in one of the largest deformation foci becomes critical, which leads to formation of a crack and fracture of the sample (figure 7, pattern 5).

For the irradiated alloys, deformation diagrams for maximum values of vertical $\sigma=f(\varepsilon_{YY})$ and horizontal $\sigma=f(\varepsilon_{XX}$ deformations inside plastic deformation foci on the surface of the samples (figure 6 *c*, *d*) have been plotted. It can be seen that the initial stage includes a linear dependency $\sigma=f(\varepsilon_{YY})$ which ends at a voltage of ~35 MPa. However, in the non-irradiated sample, the linear dependence region ends at lower values of ~ 18 MPa. The end of the linear region of the dependence on the curve $\sigma=f(\varepsilon_{YY})$ correlates with the end of the vertical region on the dependence $\sigma=f(\varepsilon_{XX})$ (point B, figure 6 *c*, *d*). The functional dependence close to parabolic begins with a voltage of ~ 35 MPa on the curve $\sigma=f(\varepsilon_{YY})$ (point B).

The maximum deformation in the vertical direction in the irradiated alloy in the plastic deformation foci before fracture is 2 times higher than the average deformation in the sample. The maximum deformation in the horizontal direction in plastic deformation foci before fracture is close to the average deformation in the sample (figure 3 and figure 6 *c, d*). The fracture process includes sample unloading, and the maximum deformation in the deformation foci on the sample surface in the horizontal direction ε_{XX} considerably decreases (figure 6, *d*, point F).

Thus, for the irradiated alloy, in contrast to the non-irradiated, similar functional dependences $\sigma=f(\varepsilon_{YY})$ and $\sigma=f(\varepsilon_{XX})$ of deformations in plastic deformation foci on the sample surface have been established depending on the direction of deformation.

4. Conclusion

Three deformation stages have been revealed on tensile diagrams $\sigma=f(\varepsilon)$ of the non-irradiated hypoeutectic silumin ASTMB179-92A. The elastic deformation stage (stage I) ends at deformation of ~0.09%, transition from stage II to III at deformation of ~0.14%, sample fracture occurs at $\varepsilon_{time.}$~0,16%. Transition from stage I to II on the deformation curve is manifested in the fact that deformation distribution patterns, in the vertical direction in stage II, almost completely lack plastic deformation foci, in which $\varepsilon_{YY}<0$, and the plastic deformation foci, in which $\varepsilon_{YY}>0$, remain. Whereas, at stage I, plastic deformation of compression and tension does exist. A correlation in the evolution of distribution patterns of horizontal and vertical relative deformations on the surface with stages in the deformation dependence $\sigma=f(\varepsilon)$ under tension of non-irradiated samples has been established. The local deformation inside plastic deformation foci in the vertical direction at the pre-fracture stage is 3 times higher than the average deformation over the entire sample. Irradiation of the surface of hypoeutectic silumin ASTMB179-92A samples results in a twofold increase in the magnitude of the fracture stress of the sample, as compared to the unirradiated sample. Investigations of deformation diagrams $\sigma=f(\varepsilon)$ of irradiated hypoeutectic ASTMB179-92A silumin under tension has not allowed to single out stages. This type of deformation curve correlates with distribution of strain fields on the sample surface: distribution patterns of relative deformations on the surface of the irradiated sample are similar to patterns when the sample is deformed from 0.04% to 0.28%.

Acknowledgement

The research was financially supported by the RSF (project No 19-79-10059).

5. References

1. Fridlander I N 2006 *Memories on the development of aerospace and nuclear technology from aluminum alloys* (Moscow: Nauka).
2. Ivanov Yu F 2017 *Russian Physics Journal* **60** No. 1 175-81.
3. Ustinov A M et al 2018 *Materials Science Forum* **938** 62-9.
4. Kopanitsa D G et al 2018 *Russian Physics Journal* **60** No. 9 177-85.
5. Jones R, Wykes C 1983 *Hohgraphic and Speckle Interferometty* (Cambridge: University' Press).
6. Koval N N and Ivanov Yu F 2008 *Russian Physics Journal* **51** No. 5 60-70.

7. Manalo A et al 2016 *Construction and Building Materials* **105** 365–76.
8. Vildeman V E et al 2017 *Mechanics of Solids* **52** No 5 488-94.
9. Panin V E 1998 *Theor. & Appl. Fract. Mech.* **30**. No1 1-12.
10. Derugin E E 2003 *Fatique Fract. Engng Mater. Struct.* **26** 295-304.
11. Honeycombe R W K 1984 *The plastic deformation of metals.* Edward Arnold. 1984, 483 p.
12. Polukhin P I, Gorelik S S and Vorontsov V K 1982 *Physical basis of plastic deformation* (Moscow: Metallurgy).

IOP Conf. Series: Materials Science and Engineering **665** (2019) 012006 doi:10.1088/1757-899X/665/1/012006

Monitoring of properties of epoxy molding compounds used in electronics for protection and hermetic sealing of microcircuits

A N Kokatev, N M Iakovleva, K V Stepanova, N Yu Ershova, K V Belov

Department of information-measuring system and physical electronics, Institute of physics and technology, Petrozavodsk State University (PetrSU), Lenin Str., 33, 185910, Petrozavodsk, Russia

e-mail: nelan-oksid@bk.ru

Abstract. It is known that epoxy molding compounds have a significant impact on productivity, service life and reliability of electronic microcircuit components. Wrong choice of a molding mixture can lead to mechanical stresses, deformation, cracking or lamination of the protective packaging followed by microcircuit breakage. The study summarizes information on composition, main operating characteristics and application methods of epoxy molding compounds as well as their impact on microcircuit shape distortion. Main methods for control of physical and mechanical properties of epoxy molding compounds with the most significant impact on microcircuit hermetic sealing and on formation of different kinds of defects in the protective packing.

1. Introduction

For correct and long service of electronic microcircuit components in automobile, aviation, ship, rocket and space equipment it is often required to protect the components from the environment, including mechanical (shocks, vibrations), chemical (moisture, heat, UV) and electrical impact. A solution to this problem is to apply various sealing coatings (glass, ceramic, silicone, epoxy, etc.) which provide for protection and heat dissipation as well as simplified installation of electronic components on a printed circuit board. The most popular type of protective sealing coatings are epoxy molding compounds (EMC) because of their affordability and advantages related to the productivity of the hermetic sealing process.

Literature analysis showed that by now a huge amount of theoretical and empirical data has been accumulated on peculiarities of electronic microcircuit components hermetic sealing with epoxy molding compounds [1, 2]. Though in practice manufacturers often observe certain problems during the hermetic sealing process, e.g. deformation, cracking or lamination of protective packing surface, formation of defects, cavities and discontinuities [1-5]. Understanding the impact of the EMC composition and properties on mechanical stresses and deformation of protective packing and substrate at the final stage of microcircuit manufacturing process is one of the most important problems for microelectronics.

This study is devoted to generalization of data on EMC composition and properties and on reasons for defects formation and protective sealing shape distortion during hermetic sealing process. In this study results of research on EMC composition, properties and application procedures are described with a purpose to solve this problem.

2. Basic information about EMC

2.1 Ingredient composition of EMC and its role

Nowadays there are plenty of MC of different shapes, weight, color and properties developed, manufactured and used for hermetic sealing of electronic components of devices in micro- and nanoelectronics. The main manufacturers of EMC are the following: Henkel Electronic Materials LLC, Hitachi Chemical Co.Ltd., Kyocera Corporation, Nepes AMC, Skymart Technologies PTE LTD, Sumitomo Bakelite Co. Ltd, Eler, KCC Corporation. Every company has a product range consisting of several dozens of EMC types. This variety is caused by a wide spectrum of tasks solved by a manufacturer.

EMC are thermoactive composite materials consisting of epoxy resins, hardeners, a filler (SiO_2 or Al_2O_3), catalysts, coloring agents, coupling and releasing agents (see figure 1a) [1]. During manufacture primary components of EMC are mixed at room temperature then transferred to a kneader, heated up to some particular temperature and then mixed until obtaining a homogeneous mixture. After mixing the material is rolled in sheets and cooled. Obtained sheet are crushed into powder, then the powder is pelleted (see figure 1b), the pellets are used for hermetic sealing of microcircuits [6].

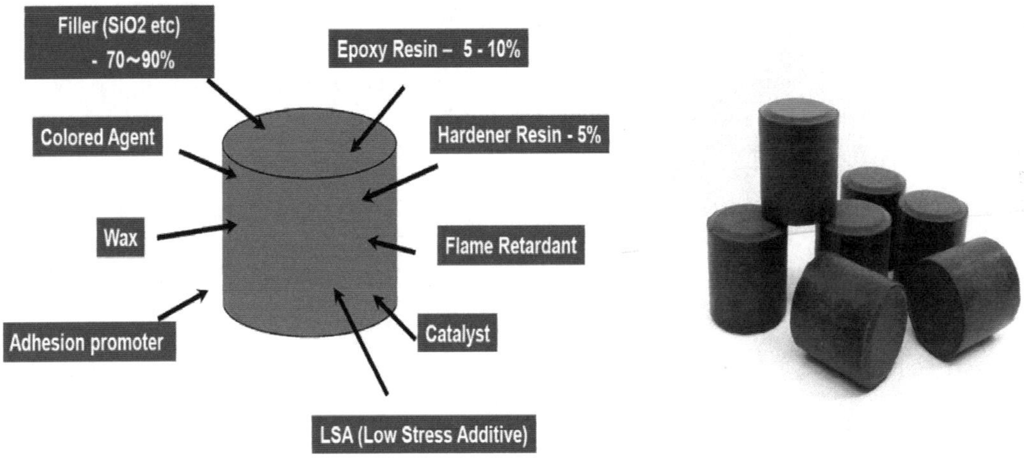

(a) (b)

Figure 1. Typical ingredient composition of EMC (a), samples of EMC in pellet forms (b)

The main component of any molding mixture is a resin or a mixture of resins. The epoxy resin is the most popular one, since it has well-balanced physical and chemical properties (high adhesion strength, chemical strength, thermal resistance and moisture resistance, low viscosity and shrinkage ratio), low cost and good aesthetic properties. Besides use of the epoxy resin is preferable with regard to the hermetic sealing process, because it has a low glass transition temperature and short gel time [1, 6].

Phenolformaldehyde resins, anhydrides, amines or aliphatic alcohols can be used as a hardener. Anhydrides are usually used to produce heat-proof EMC and phenolic compounds increase moisture resistance of a coating. Every hardener has its own advantages and disadvantages, e.g. price or productivity, this is why it is chosen according to properties of the hardened coating [6].

Catalysts are used as a part of EMC to speed up the hardening process. Wrong choice of a catalyst can make EMC less stable and decrease the service life.

Use of different hardeners, catalysts and variation of the quantitative ratio of them and epoxy resins allows to control mechanical and adhesion strength, flow characteristics, thermal expansion coefficient, moisture resistance, glass transition temperature, specific resistivity, dielectric properties and service life of EMC.

Filler type also has a significant impact on final properties of EMC. For example, if Al_2O_3 is used as a part of EMC it increases thermal conductivity and changes magnetic properties and if addition of

SiO_2 increases strength and electric insulation properties. Correct choice of a filler and its content in EMC allows to solve most of the problems occurring during hermetic sealing. It is critical that the filler and sealed substrate, which usually consists of several inhomogeneous materials - a printed circuit board, glue, a semiconductor element, had close values of thermal expansion coefficient. It allows to prevent substrate deformation and EMC delamination from it. Such filler parameters as shape, diameter, size distribution and specific surface area of particles have impact on the packing density, a well a on flow characteristics or EMC ability to fill cavities. When EMC is in a molten state the filler with particles of spherical shape (amorphous SiO_2) has lower viscosity and flows better. On the other hand a filler with particles of plate-like shape (crystalline SiO_2, Al_2O_3 or Si_3N_4) has higher viscosity because of bigger particles and is used to manufacture EMC with high thermal conductivity coefficient. This is why size and shape of filler particles are among the most important parameters for choosing EMC [1, 2, 6].

Another components affecting EMC properties are silane coupling agents and colorants. The former are used to strengthen the bonding between an inorganic filler and an organic resin matrix as well as to improve EMC adhesion with the substrate and semiconductor elements, the latter serve indication purposes and simplify the laser marking process. [6].

Thus, components contained in EMC provide for required physical and chemical properties, adhesion strength and mold-ability. Percentage ratio of components and the chemical composition of EMC itself can differ depending on required operating properties of a protective sealing, type of the substrate used and application method.

2.2 Main EMC properties and their influences on packaging

All EMC manufactured are characterized by a et of properties and parameters which define their potential fields of application. All EMC properties can be divided into four groups: (1) industrial, (2) hygro-thermomechanical, (3) electrical and (4) chemical properties. Several industrial and hygro-thermomechanical properties are crucial for selection of EMC. They define EMC applicability for particular processes: (1) which hermetic sealing method shall be used, (2) which electronic microcircuit elements can be sealed, (3) shape and dimensions of obtained protective packaging. These properties include the following: spiral flow, mold shrinkage, thermal expansion coefficient, glass transition temperature and gelation time [2].

The spiral flow is one of the most important EMC characteristics, since it has a direct impact on the behavior of the molten molding mixture in a mold die. This parameter is used to compare different materials, to evaluate EMC quality and to determine compatibility of EMC with a particular mold die. The spiral flow on its own is not a measure of viscosity even though it can be used for its evaluation implicitly. The spiral flow is a characteristic of a flow and shows the distance which molten EMC will pass in a mold die under certain temperature and pressure [2, 7].

The mold shrinkage is a value indicating the difference in linear dimensions of a m and a sample extracted from it at room temperature [2, 7]. It depends directly on the composition (type of resin, filler, reinforcing elements, etc.) and thermal expansion coefficient of EMC, which can be chosen in such a way that the size difference is minimum. Modern EMC have shrinkage ratio of 0.01 to 0.3%. EMC shall be elected in such a way that its shrinkage ratio corresponds to the shrinkage ratio of the substrate material. It ensures stability of size and shape of materials after hermetic sealing and prevents occurring of different types of stresses leading to deformation of the substrate with EMC.

The glass transition temperature (T_g) is another important property of any epoxy resin. It is the temperature at which thermoactive EMC transit from a solid (glass state) state into a softer one (elastic) [2, 7]. The transition of EMC from glass to elastic (rubberlike) state leads to a sharp increase (by 2-3 times) of the thermal expansion coefficient and decrease of hardness and modulus of elasticity. The glass transition temperature shall not be confused with the melting point. It must be pointed out that actually T_g is not a particular temperature, but a temperature range where the mobility of polymeric chains increases significantly which allows EMC to flow along the mold die filling cavities. The T_g value can vary depending on EMC degree of cure, this is why in is crucial to maintain storage temperature at all times. Besides the glass transition temperature (T_g) of EMC can vary because of moisture adsorption, which also shall be taken into account.

The thermal expansion coefficient (TEC) is a parameter characterizing relative change of volume or linear dimensions of EMC in case of temperature increase by 1°C at constant pressure [2, 7]. Has a dimension of reciprocal temperature. In practice EMC are characterized by two different TEC values, which are usually called ά1 and ά2. Value ά1 is TEC below the glass transition temperature and ά2 - above T_g. If only one TEC value is given in EMC specification, then most probably it is ά1. Since different materials expand differently at the same heating rate, then for hermetic sealing one shall choose EMC with a TEC closest in value to TEC of the substrate. It allows to reduce the risk of protective packaging shape distortion and EMC lamination.

The gelation time is the time required by EMC for transition from a liquid state to a solid (gel-like) state at the hermetic sealing temperature [2, 7]. When this time expires EMC stops flowing and the process of mold die filling stops. Decrease in the gelation time leads to reduction of time required to fill the mold die, which requires pressure and EMC flow rate increase. EMC flow rate increase can result in damaging microcircuit sealing elements, e.g. displacement or tearing of wire contacts. Otherwise if low flow rate is used for EMC with short gelation time, it can result in the formation of cavities and incomplete sealing of the microcircuit. The gelation time can also be a reason for cracks formation and weak adhesion of EMC to the substrate material.

Thus EMC properties are crucial when choosing the method and parameters for hermetic sealing, protective packaging structure and field of application in electronics.

3. Epoxy molding compounds application methods

The following two methods are commonly used for sealing of microcircuits: transfer or pressurized casting method and compression method [1, 2, 7].

The pressurized casting method is a process when molten EMC is brought into a closed mold die at particular temperature and pressure (see figure 2a). The compression method is a method when particular amount of molten EMC is put into the bottom part of a heated mold die. After that the mold die is closed with the top movable part to which a microcircuit is fixed. In the course of closing the mold die is pressurized, which results in EMC filling all cavities and getting in contact with the microcircuit (see figure 2b). During hermetic sealing temperature and pressure are maintained at a constant level until full cure of EMC.

Recently the compression method with EMC powder gradually displaces the transfer method from the market, even though the latter had been the main hermetic sealing method for several decades.

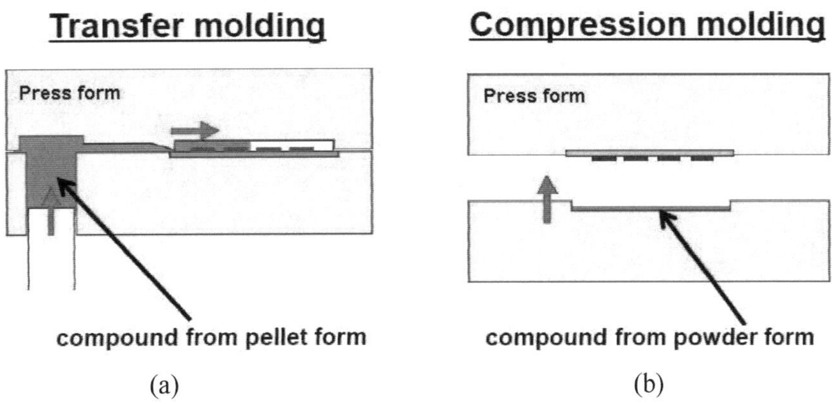

Figure 2. Schematic diagram of microcircuit hermetic sealing process by means of transfer method (a) and compression method (b)

This may be explained by the fact that in case the transfer method is used there is a high probability of defects , loss of integrity and deformation of the protective packaging because of many impact factors (pressure, temperature, EMC composition and properties) affecting the process of filling the mold die cavities and subsequent EMC curing. Also the compression method has several advantages [1]:

1. During hermetic sealing process one can control the pressure value which allows to reduce the number of cavities in the protective packaging. The possibility to decrease the pressure during hermetic sealing allows to apply this method for hermetic sealing of devices sensitive to pressure, e,g, MEMS devices.
2. Low equipment maintenance costs. A special film is used for mold die protection, which allows to perform continuous hermetic sealing process and increase the mold die service life.
3. Allows to change protective packaging thickness. Can be applied to obtain very thin protective coatings, hermetic sealing of multi-chip modules and hybrid microcircuits.
4. Cost effectiveness. It is possible to control EMC expenditure during hermetic sealing process.
5. Short length of the EMC path in the mold die, which reduces the probability of filler segregation.

Based on the above said the compression method has obvious advantages and allows to solve more industrial tasks when compared to the transfer method. But it is necessary to pay due regard to EMC composition and properties when using this method.

4. Results and Discussion

It has been established that the main physical and mechanical EMC properties with an impact on generation of various types of defects in the protective packing during molding are the following: spiral flow, mold shrinkage, thermal expansion coefficient, glass transition temperature and gelation time [1-3, 7-9]. Right choice of molding mixture allows to prevent problems related to production effectiveness, stress and deformation of the protective sealing, cracking and lamination of EMC surface (see figure 3). Though it is not always possible in practice. The main problem is that molding compounds' manufacturers often specify only approximate values of the EMC parameters or do not give them at all. One can assume that the most probable reasons for that are the following: (a) - change of EMC properties in time even when storage conditions are fulfilled [9]; (b) - measuring conditions used by EMC manufacturers, e.g. for spiral flow, can be different from microcircuit sealing conditions at particular manufacturing site; (c) - a manufacturer can use some internal standards to determine EMC properties and these standards can differ from commonly used.

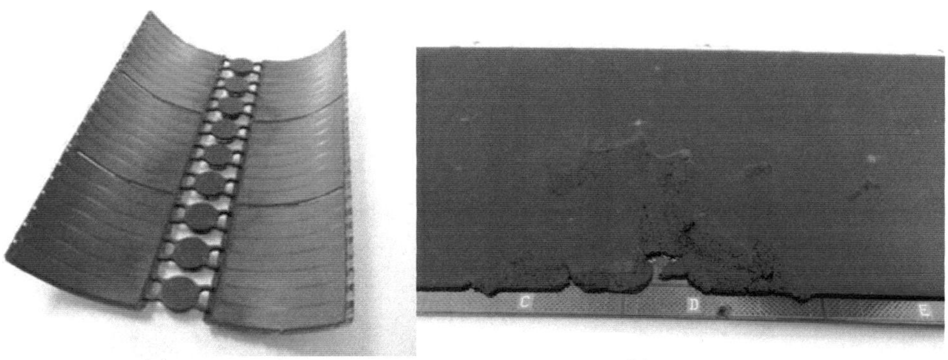

(a) (b)

Figure 3. Photos of deformated substrate - (a) and EMC defects (faults) appearing in the course of the sealing process - (b).

It was established that at the present time there are several ways to solve this problem: (1) using EMC with higher glass transition temperature (T_g) for microcircuit hermetic sealing [1], which improves stability of substrate size at different temperatures. Although using EMC with high T_g puts serious restrictions on using particular equipment, choice of substrate material and hermetic sealing parameters; (2) repetitive optimization of the manufacturing process by selection of hermetic sealing parameters (temperature, pressure, etc.). But in practice this is a very expensive (heavy expenditure of microcircuits and EMC) and time-consuming process; (3) preliminary modeling of hermetic sealing process using special software for casting process simulation, such as ANSYS or Moldex3D [10].

A distinguishing feature of computer modeling is accuracy and efficiency of optimization of the hermetic sealing process in the context of selection of temperature, pressure and mold die filling rate, protective packaging deformation reduction and EMC shrinkage. But to get correct modeling results one shall know exact value of spiral flow, mold shrinkage coefficient, thermal expansion coefficient, glass transition temperature and gelation time of EMC.The study conducted identified a set of methods and standards for measuring and control of EMC physical and mechanical properties required to model the microcircuit sealing process (table 1).

Table 1. EMC properties control methods

Parameter	Measuring units	Method	Standard
Spiral flow	inch	Measuring the spiral flow	ASTM D3123 [11], SEMI G11-88 [12]
Mold shrinkage	%	Thermomechanical analysis	ASTM D955-08 [15] ISO 2577 [16]
Thermal expansion coefficient ($\alpha_{1,2}$)	$x10^{-5}/°C$	Thermomechanical analysis	ASTM D696 [13], SEMI G13-88 [14]
Glass transition temperature (T_g)	°C	Thermomechanical analysis	ASTM D696 [13], SEMI G13-88 [14]
Gelation time	s	Change of EMC polymerization rate	SEMI G11-88 [12]

It is worth mentioning that EMC properties control shall better be performed at the production floor. In practice most companies are forced to deviate from the standards specified in Table 1 or to develop their own methods for EMC properties assessment. The reason for this is the necessity of adapting testing conditions to conditions and requirements of a particular production site. The critical factor of EMC properties production control is testing which allows to obtain relevant information which can be useful for production department and modeling department.

5. Conclusion

The study shows that the incoming inspection of EMC properties is the most efficient and cost-effective way to minimize the risks related to mechanical stresses, deformation or loss of integrity of protective packing in course of hermetic sealing of electronic microcircuit components. The results obtained can be used by both scientists and engineers of companies involved in electronic microcircuit components manufacture for correct selection of EMC, optimization of manufacturing processes, studying and understanding of causes for various types of deformation and stresses during hermetic sealing.

References

[1] Lu D and Wong C P 2017 *Materials for Advanced Packaging. Second edition.* (Cham: Springer)
[2] Ardebili H and Pecht M G 2009 *Encapsulation Technologies for Electronic Applications* (Burlington, USA: Elsevier)
[3] Mokhtar U et al 2008 *J. Solid State Sci. Technol.* **16(2)** 189-97
[4] Tsai M-Y, Wang C T and Hsu C H 2006 *IEEE Trans. Adv. Packag.* **29(3)** 625-35

[5] Lakhera N et al 2015 *Int. Symp. on Microelectronics: FALL 2015.* No 1 (Orlando) pp 000696-701.
[6] Glover J 1996 *Technical paper: Formulating Coating Powders for the Passive Electronic Component Industry* (New York: Dexter Electronic Materials) p 4
[7] Tummala R R, Rymaszewski E J and Klopfenstein A G 1997 *Microelectronics Packaging Handbook. Semiconductor Packaging.* (Boston: Kluwer Academic Publisher)
[8] Kim Y, Lee H, Zhang X and Park S 2014 *IEEE Transactions on CPMT.* **4(10)** 1589-97
[9] Tachapitunsuk J, Ugsornrat K et al 2017 IOP *Conf. Series: J. Phys.* **901** 012090
[10] Turng L S and Wang V W 1993 *J. Reinf. Plast. Compos.* **12** 506-19
[11] ASTM D3123 1998 *Standard test method for spiral flow of low-pressure thermosetting molding compounds* (American Society for Testing and Materials)
[12] SEMI G11-88 1988 *Recommended practice for RAM follower gel time and spiral flow of thermal setting molding compounds* (SEMI standards)
[13] ASTM D696 2003 *Standard test method for coefficient of linear thermal expansion of plastics between -30°C and 30°C with a vitreous silica dilatometer* (American Society for Testing and Materials)
[14] SEMI G13-88 (Reapproved 0811) 1988 *Standard Test Method for Expansion Characteristics of Molding Compounds* (SEMI standards)
[15] ASTM D955 – 08 2014 *Standard Test Method of Measuring Shrinkage from Mold Dimensions of Thermoplastics* (American Society for Testing and Materials)
[16] ISO 2577 2007 *Plastics Thermosetting moulding materials Determination of shrinkage* (International Organization for Standardization)

Acknowledgments
The authors acknowledge the Ministry of Science and Higher Education of the Russian Federation for the financial support under the project no. 05.577.21.0293, unique identifier RFMEFI57718X0293.

MSMEE 2019 IOP Publishing
IOP Conf. Series: Materials Science and Engineering **665** (2019) 012007 doi:10.1088/1757-899X/665/1/012007

Decision making system for industrial companies

A O Loyko[1, a], S A Gusev[2, b]

[1,2] Moscow Finance and Law University 8 Presnenskaya Embankment, Capital City Tower, 9th Floor, Moscow 123317 Russia

[a] loykoanastasia@gmail.com, [b] fulcrum@mail.ru

Abstract. In a competitive business, successful transactions play a significant role of industrial enterprises' management. The decision support system that is offered by the author provides using the diagnosis of supplier, which is based on the financial statements by calculating financial ratios and assessing the dynamics of changes values of ratios and indicators of financial statements.

1.Introduction

Decision support is important problem in science and technology. This is especially true for industrial enterprises in which the following the conditions of the contract is an essential condition for the health economic.

The decision support system provides using diagnosis of supplier, which is based on the provided financial statements by calculating financial ratios and assessing the dynamics of changes values of ratios and indicators of reporting items (Figure 1).

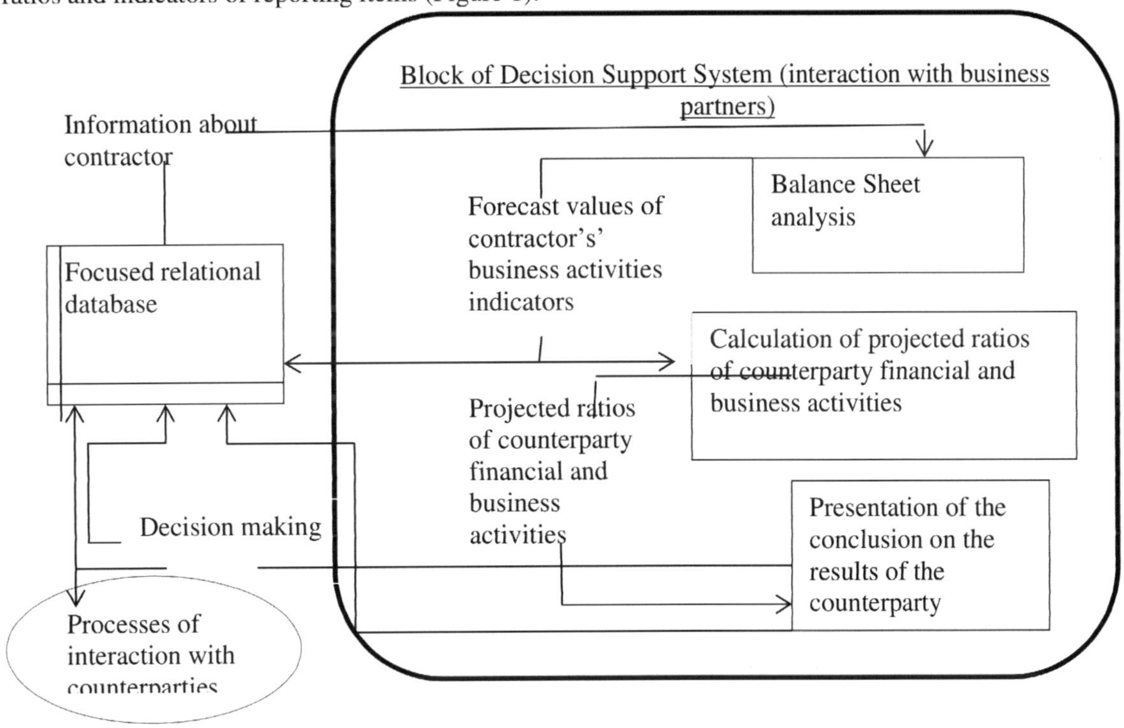

Figure 1. Scheme of Decision Support process

2.Results and Discussion

Diagnostics is based on the following documents (data): Accounting documents for the 3 (three) last reporting years and for the last reporting period, drawn up in accordance with paragraph 3 of Art. 14;

IOP Conf. Series: Materials Science and Engineering **665** (2019) 012007 doi:10.1088/1757-899X/665/1/012007

Section 1. Art. 30 of the Federal Law of 06.12.2011 №402 "O buhgalterscom ychete", p. 48 PBU 4/99: Balance Sheet; Profit and Losses Statement; Cash Flow Statement [1]

Diagnostics may be supplemented by information from the following sources:

- The current structure of shareholders (owners), indicating their names;
- The list (name) of the main service provider of banks.
- Availability of insurance policies (indicating the type of insurance, the term of the policy and the liability limit of the insurance company).
- Distribution of total revenue by types of products (works, services) and by geography of sales.
- Information is describing what proportion of sales is own-produced products and what proportion of the company's total sales is made by products produced by subcontractors. Consolidated revenue by group (if the company is in a group of related companies).
- Interpretation of receivables and (or) accounts payable with a share of details up to 10%.
- Interpretation of fixed assets' structure [2].

Table 1 is based on the above documents (data), (in the appropriate units of measurement):

Table 1. Data Form for supplier's Diagnostics

Parameters of planning deal		**value**
Contract price, thousand roubles.		
Term of agreement, years		
Common parameters		**value**
The lifetime of the Counterparty, years		
Share of the Russian Federation and organizations affiliated with the Russian Federation in the authorized capital, %		
Indicators of financial and economic activity		
Balance Sheet (form №1)	**code term of balance sheet**	**value**
Financial investments, thousand rubles	1170	
Total section I, thousand rubles	1100	
Financial investments (with the exception of cash equivalents), thousand rubles	1240	
Cash and cash equivalents, thousand rubles	1250	
Total section II, thousand rubles	1200	
Total section III, thousand rubles	1300	
Total section IV, thousand rubles	1400	
Revenues of future periods, thousand rubles	1530	
Estimated liabilities, RUB thousand	1540	
Total section V, thousand rubles	1500	
Balance, thousand rubles	1600	
Report oa financial results (form №2)	**code term of balance sheet**	**value**
Revenue, thousand rubles Profit (loss) from sales, thousand rubles	2110	
Revenue, thousand rubles Profit (loss) from sales, thousand rubles	2200	

Financial condition of the Counterparty are calculated on the contract information and reporting data, the coefficients characterizing. The ratio of the transaction volume and the net assets of the Counterparty [3]. Calculated with the data of the draft contract and financial statements (form No. 1) according to the formula: K1= Contract price/ (code 1600- (code1400+code1500-code1530))

The profitability ratio of the suplier's financial and business activities (the ratio of profit to revenue). Calculated on the basis of Balance Sheet (form No. 2) according to the formula:

K2= code 2200/code2110

Leverage ratio of the supplier (the ratio of balance to capital). Calculated on the basis of accounting data (form No. 1) according to the formula:

K3= (code 1400 + code 1500)/code1600

The absolute liquidity ratio (the ratio of cash and short-term financial investments to short-term liabilities). Calculated on the basis of Balance Sheet (form No. 1) according to the formula:

K4=(code 1240 + code 1250)/(code 1500 - code 1530 - code 1540)

Determining the total credit risk of the company. Each indicator, depending on its value, is assigned a score from 1 to 5 according to Table 2.

Table 2. Scoring indicators.

Score **General indicators**	**Score corresponding to score**				
	5	**4**	**3**	**2**	**1**
The term of the transaction, years	≥ 5 years	3 - 5 years	2 -3 years	1 - 2 years	< 1 years
Time of existence, years	≤ 1 years	1 -5 years	5 - 10 years	10 -20 years	>20 years
Share of the Russian Federation and organizations affiliated with the Russian Federation in the authorized capital,%	-	> 2%	2% -25%	25% - 50%	50 - 100 %
Indicators of financial and economic activity					
The ratio of transaction volume and net assets,% Profitability of financial and economic activity,% Leverage ratio,% Absolute liquidity ratio,%	≥100 %	100%-50%	50%-30%	30% - 10%	<10 %
The ratio of transaction volume and net assets,% Profitability of financial and economic activity,% Leverage ratio,% Absolute liquidity ratio,%	≤ 0%	0 - 4%	4% - 10%	10 - 20%	>20 %
The ratio of transaction volume and net assets,% Profitability of financial and economic activity,% Leverage ratio,% Absolute liquidity ratio,%	≤ 90%	90% - 80%	80% - 70%	70% - 60%	>60 %
The ratio of transaction volume and net assets,% Profitability of financial and economic activity,% Leverage ratio,% Absolute liquidity ratio,%	≤ 20%	20% - 30%	30% - 40%	40% - 50%	>50 %
Total					

The actual value of each indicator, as well as its corresponding score, are recorded in the columns of Table 3 "Indicator Value", "Indicator Score", respectively.

Table 3. Calculation of the total credit risk

Evaluation parameters	**Weight indicator**	**Value of the indicator**	**Score**	**Total score**
Term of agreement, years	**10%**			
General indicators	**20%**			
The lifetime of the Counterparty, years	10%			
Share of the Russian Federation and organizations affiliated with the RF in the authorized capital of the counterparty,%	10%			
Indicators of financial and economic activity	**70%**			
The ratio of transaction volume and net assets,%	17.5%			
Profitability of financial and economic activity,%	17.5%			
Leverage ratio,%	17.5%			
Absolute liquidity ratio,%	17.5%			
Total risk	**100%**			

In the column "Final Score" of table 3 indicates the point value of the indicator based on its weight, determined by the formula:

Total score of the indicator = Weight of the indicator * Score of the indicator

The final risk score is calculated by summing the calculated in accordance with the final points of each indicator: Total risk indicator = □∑ Weight of indicator * Score of indicator

The final indicator is indicated in the lower right field of Table 2.

Characteristics of the Counterparty's credit risk. The total risk index has limits from 1 to 5, which corresponds to the probability of realization of credit risk from low to very high.

The decision support system proposed by the author helps to minimize the following types of risks of industrial enterprises in the process of interaction with suppliers:

losses associated with the counterparty's failure to fulfill contractual obligations. The risk of default by banks. non-fulfillment of obligations by counterparties when providing organizations of the Group of financing non-performance by counterparties of their obligations under contracts (other than financial contracts) involving the occurrence of receivables. failure to fulfill obligations by suppliers (performers) / buyers (customers) of products (works, services)[4].

The expediency of provision and the amount of advance payment security is determined on the basis of various options proposed by the decision support system. Solutions are proposed on the basis of an assessment of the financial condition of the counterparty and the risk of non-performance of obligations by the counterparty under contracts entered into on a non-competitive basis. It also determines the volume (total value) of goods, works, services sold by the organization on terms with a deferred payment [5].

The information that is necessary to conduct a risk assessment (financial (accounting) reporting, other documents and materials) are submitted to the Responsible Division by the structural units that initiate the conclusion of the relevant contract. The Responsible Division make a decision about the expediency of the provision and the amount of the collateral for the advance payment of the contract and permissible forms of payment. It is determined on the basis of the final indicator (Table 4)

Table 4. The standard amount of the advance payment and the form of deferred payment.

Total risc	**The bank guarantee,% of the advance**		Acceptable form of payment / security
1 <= indicator < 1,5	Low	<25%	Without security, or a pledge, or a guarantee of an organization with a "Low" assessment of the probability of credit risk, or a bank guarantee of at least 25% of the amount of the delivery on credit
1,5 <= indicator < 2		>25%	A pledge or guarantee of an organization with a "Low" assessment of the probability of the realization of credit risk or a bank guarantee in the amount of at least 50% of the amount of the delivery on credit
2 <= indicator < 2,5	Middl e	>50%	Bank guarantee for 100% of the amount of delivery on credit
2,5 <= indicator < 3	High	100%	Prepayment or bank guarantee for 100% of the amount of delivery on credit
3 <= indicator < 5	Very high		Acceptable form of payment / security

3. Conclusions

In this way , based on the support system, the company decides whether or not to provide collateral for the advance / deferred payment under the contract and the form and amount of the security (if it is reasonable to provide collateral) and forms the terms of the contracts in terms of advance / installment

IOP Conf. Series: Materials Science and Engineering **665** (2019) 012007 doi:10.1088/1757-899X/665/1/012007

payments, the amount of security of receivables. This article is concerning the problem in choosing the most stable business partner. Taking into the consideration the specific function of industrial enterprises some terms of the contract may lead to the failure of all process. The decision support system offered by the author is economically justified.

References

[1] Lindström B, Golkar A, Jangard S, Philippe N, Olsson A 2019 *PNAS* **116(10)** 4732
[2] Arnott D, Pervan G 2012 *JAIS* **13** 11
[3] Andreichikov A, Dekatov D, Kremenov S 2010 *M Information Systems Management* **25** 149
[4] Feng N, Arnott D, Gao S 2018 *VJornal of Decision Systems* **5** 23

Aluminate spinel-based SHS pigments for fire-resistant protective and decorative coatings

N I Radishevskaya, A Yu Nazarova, N G Kasatsky

Tomsk Scientific Centre of the Siberian Branch of the Russian Academy of Sciences, 10/3, Akademicheskii Pr., Tomsk, 634055, Russia

e-mail: osm.nazarova@yandex.ru

Abstract. In the work, color spinel-based inorganic pigments obtained by the SHS method are used in fire-resistant protective and decorative aluminophosphate-bonded coatings. X-ray diffraction (DRON-UM1 diffractometer, filtered CuKα radiation) and IR spectroscopic (Nicolet 5700 FTIR spectrometer) analyzes revealed that the coating contained $AlPO_4$, $Al_2(HPO_3)_3$ and $Al(H_2PO_4)_3$ phases and spinel. Thermal analysis performed on an SDT Q600 thermal analyzer showed a slight mass loss related to the polymerization of phosphates and the formation of $P_2O_7^{2-}$ anion in the range of 500-550 °C, which was also confirmed by IR spectroscopy. The color of protective and decorative coatings is maintained at temperatures of up to 1000 °C. Optical studies (Axiovert 200M) showed that color coatings subjected to heat treatment at 300 °C have a uniform structure that does not contain cracks, and therefore they can be used for metal, concrete and brick surfaces.

1. Introduction

The creation of various composite materials increases the demand for high-quality pigments that are resistant to UV irradiation, elevated temperatures and corrosive media, which in turn stimulates the expansion of the color palette and the methods for obtaining inorganic pigments. Spinelids are considered to be the best ceramic pigments [Masl.]. They are characterized by high-temperature formation conditions.

Large volumes of spinelides are obtained, as a rule, by the ceramic method, which includes annealing the corresponding oxides at temperatures of 1200–1400 °C, and by the sol-gel method followed by calcination [1–5]. All these methods are expensive due to increased energy consumption, therefore, the reduction in the cost of spinels through the use of the energy efficient method of self-propagating high-temperature synthesis (SHS) is an urgent problem. Aluminum powders (ASD-4 and ASD-6) produced in sufficient volumes by metallurgical enterprises were used to develop the SHS procedure for obtaining aluminum spinels [6]. The high speed of processes at which the temperatures required for the synthesis of spinels are reached, simple equipment and waste-free production are also the advantages of this method.

Spinels are widely used in the ceramic industry for the production of ceramics, refractory materials; color spinels are contained in heat-resistant paints. Spinel SHS pigments with increased chemical resistance to mineral acids, high melting temperatures, water resistance, Mohs' hardness of 7.5-8 and the ability not to fade in the sun can be used in white and color refractory protective and decorative aluminophosphate-bonded coatings. Aluminophosphate binder (APB) is a product of the interaction of aluminum hydroxide with orthophosphoric acid and possesses good binding properties due to the formation of aluminum phosphates [7–10]. Refractory SHS spinelides added to APB were used to obtain aluminophosphate compositions with refractory and decorative properties.

The purpose of this work was to obtain protective and decorative aluminophosphate-bonded coatings using spinel SHS pigments and study the phase composition and heat resistance of the compositions obtained.

2. Experimental procedure

Pigments were synthesized from ZnO, MgO, CoO, Co_2O_3, Co_3O_4, Al_2O_3 oxides (high grade, pure grade) and (or) cobalt salts, and aluminum powder Al (ASD-4) was used as a fuel. The synthesis was conducted in the layer-by-layer combustion mode at atmospheric pressure in air in the SHS setup.

APB was prepared from a mixture of orthophosphoric acid and aluminum hydroxide with a ratio of H_3PO_4 :$Al(OH)_3$ = 100 ml: 22.5 g by heating to temperatures of 80 ÷ 100 °C with constant stirring.

Spinel-based SHS pigments were introduced into aluminophosphate binder in the amount of 10 -15 mas. % with a small amount of boric acid.

The microstructure of the protective and decorative coating was investigated by optical microscopy (Axiovert 200M). Obtained pigments and composite decorative material were examined using X-ray diffraction analysis (diffractometer DRON-UM1, filtered CuKα radiation) and IR spectroscopy (Nicolet 5700 FTIR spectrometer with a diffuse reflectance attachment, KBr). Complex thermal analysis of a ceramic coating consisting of APB and spinel-based inorganic pigment was performed using an SDT Q600 thermal analyzer. The particle size analysis of SHS pigments was performed on a Mastersizer-2000 analyzer.

3. Discussion of results

Pigments were synthesized at atmospheric pressure in air in the mode of layer-by-layer combustion and natural filtration.

The main reactions that provide the SHS process are exothermic oxidation of aluminum. For white magnesium aluminate spinel and light blue pigments based on it, the leading reaction is direct aluminum oxidation during which a large amount of heat is released [11, 12]:

$$4Al+3O_2=2Al_2O_3+3350\text{ kJ} \quad (1)$$

For pigments based on cobalt oxides, aluminothermic reactions develop along with direct oxidation of aluminum:

$$CoO+2Al+O_2=Al_2O_3+Co \quad (2)$$

$$2Co_2O_3+8Al+3O_2=4Co+4Al_2O_3 \quad (3)$$

As a result, a reaction mixture is heated to temperatures above 1000 °C, after which the synthesis of spinels begins, proceeding also with the release of heat:

$$MeO+Al_2O_3=MeAl_2O_4, \quad \text{where Me-Mg, Co, Zn} \quad (4)$$

The last stage of SHS is volume combustion related to oxidation of cobalt obtained during thermite reactions, its interaction with aluminum oxide, polymorphic transformations of Al_2O_3, etc.

The advantage of SHS of spinel pigments is their finely dispersed state obtained directly in the combustion wave. The particle size analysis of SHS pigments showed that the average particle size was 4–9 μm.

Pigments synthesized from magnesium aluminate spinel and cobalt chloride (MgO-Al_2O_3, MgO-CoO-Al_2O_3 system) do not contain metallic particles, and systems obtained from cobalt oxides include Co particles observed under the microscope.

Table 1 demonstrates the main and impurity compositions of synthesized medium blue pigments according to X-ray diffraction analysis.

Table 1. Phase composition of spinel-containing pigments obtained by SHS

№	System	Main phases	Impurity phases	
			Oxides	Metal
1	MgO-Al_2O_3 (white)	$MgAl_2O_4$	α-Al_2O_3, $AlBO_3$	-

IOP Conf. Series: Materials Science and Engineering **665** (2019) 012008 doi:10.1088/1757-899X/665/1/012008

2	MgO-CoO-Al_2O_3 (light blue)	$MgAl_2O_4$	α-Al_2O_3, $CoAl_2O_4$, $AlBO_3$	-
3	CoO-Al_2O_3 (blue)	$CoAl_2O_4$	α-Al_2O_3, $AlCo_2O_4$, CoO	Co
4	ZnO-MgO-CoO-Al_2O_3 (ultramarine)	$Zn_xMg_{1-x}Co_{1-x-y}Al_2O_4$	α-Al_2O_3, Al_2O_4, Θ-Al_2O_3, CoO, δ- Al_2O_3	Co

Synthesized SHS-pigments were used to obtain reliable protective and decorative aluminophosphate-bonded coatings. The optimal amount of pigment introduced into aluminophosphate binder was 10–15 wt.% with a small amount of boric acid added to slow the rate of crystallization for the technologically necessary time. Obtained paint was applied to the surface and dried. The reaction of the interaction is given by:

$$Al(OH)_3+H_3PO_4=AlPO_4+3H_2O \quad (5)$$

To create a solid ceramic coating, the surface of the product covered with dried paint was exposed to heat treatment (~ 300 °C) for 1 ÷ 2 minutes. Such temperature treatment provides water resistance to coatings [7].

Figure 1 shows the X-ray diffraction pattern of the coating based on APB and aluminum-cobalt pigment (obtained in the ZnO-MgO-CoO-Al_2O_3 system). The obtained coating is a mixture of $AlPO_4$ aluminum phosphates, crystallized in a hexagonal and orthorhombic crystal systems, and $CoAl_2O_4$ spinel.

Ceramic coating is resistant to high temperatures. Figure 2 shows the complex thermal analysis of the coating consisting of aluminophosphate binder and $CoAl_2O_4$- based inorganic pigment. Figure shows a loss of adsorbed and crystallization water with a maximum at 66.9 °C, .176.0 °C and 231.0 °C. In the range of 500-550 °C with a maximum at 538.2 °C, a slight mass loss related to the polymerization of phosphates and the formation of $P_2O_7^{2-}$ anion is observed, which is also confirmed by IR spectroscopy (Figure 3).

$$2PO_4^{2-} \rightarrow P_2O_7^{2-}+1/2O_2\uparrow \quad (6)$$

The coating is stable up to ~ 1000 °C. Spinels with high stability during heating do not react with phosphoric acid, so the color of the coating is maintained.

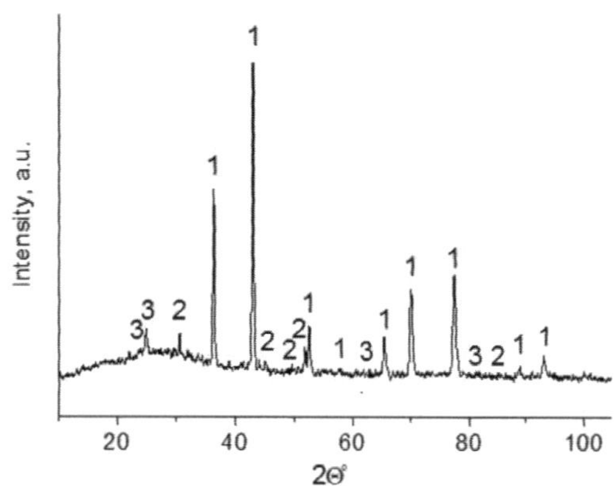

Figure 1. X-ray diffraction pattern of the ceramic coating consisting of APB and $CoAl_2O_4$ -based inorganic pigment, where 1-$CoAl_2O_4$, 2-$AlPO_4$ (hexagonal), 3-$AlPO_4$ (orthorhombic).

IR spectroscopic analysis performed on the Nicolet 5700 spectrometer showed the presence of absorption bands typical for PO_4^{3-} (932,2 cm^{-1}), HPO_4^{2-} (980,4 cm^{-1}) and $H_2PO_4^-$ (1095,7 cm^{-1}) anions in the range of 1400–900 cm^{-1}. The vibrations of bonds of tetrahedrally coordinated cobalt [CoO_4] at 687.5 cm^{-1} and octahedrally coordinated aluminum [AlO_6] at 552.3 cm^{-1} in aluminum-cobalt spinel overlap with lines of aluminum phosphate (Figure 3, curves 1 and 2).

Heat treatment of the coating at 1000 °C leads to the occurance of an absorption band at 732.3 cm^{-1}, which related to the vibration of ν(-O-) bond characteristic for $P_2O_7^{2-}$ polyanion [13]. In all cases, the absorption band at 497.6 cm^{-1} related to $H_2PO_4^-$ anion was observed. Deformation vibration of bond δ(Co ... O-P) is detected at 456 cm^{-1}, which enhances with an increase in the thermal treatment of the coating (Figure 3, curves 3 and 4). The ν(Co – O) absorption band occurs at 669.5 cm^{-1} at 1000 °C. The peak with a maximum of 1229.5 cm^{-1} is related to the valence vibration of bond ν(P = O). With an increase in the treatment temperature, the content of $AlPO_4$ in the coating increases, and the vibration of PO_4^{3-} anion contained in the composition of the coating is observed at 565.3 cm^{-1} (Figure 3, curve 4) [14].

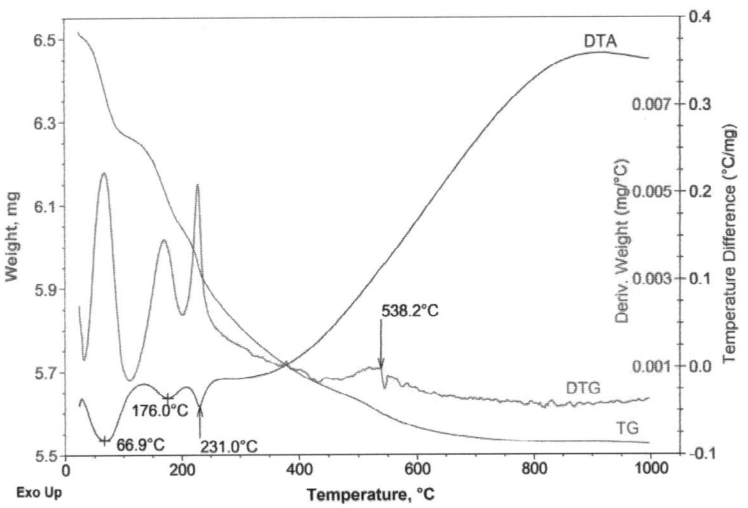

Figure 2. TG, DTG, DTA thermoanalytical curves of the composition from $CoAl_2O_4$-based inorganic pigment and aluminophosphate binder (maximum mass loss on the DTG curve are directed upwards).

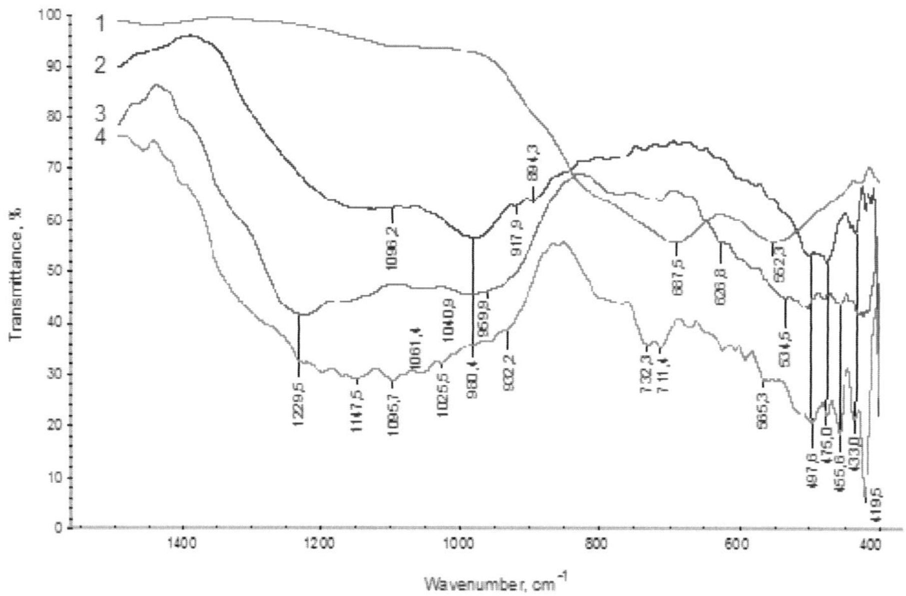

Figure 3. IR spectra of protective and decorative coatings, where 1 is pigment based on spinel $CoAl_2O_4$, 2 is APB + pigment (after drying), 3 is APB + pigment (after heat treatment at 300 °C), 4 is APB + pigment (after heat treatment at 1000 °C).

Figure 4 shows a sample of a protective-decorative coating deposited on a brick after heat treatment at ~ 300 °C.

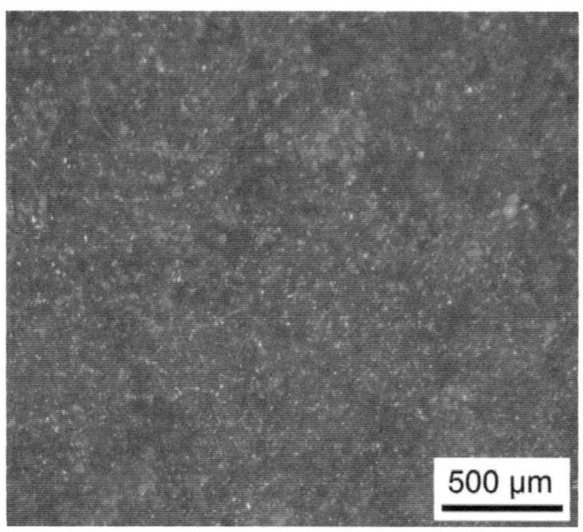

Figure 4. Microstructure of the surface of color protective and decorative ceramic coatings based on aluminophosphate binder and $CoAl_2O_4$ -based inorganic pigment deposited on brick, Axiovert 200M.

Optical studies (Axiovert 200M) showed (Fig. 4) that after heat treatment at 300 °C, the color coating has a uniform structure, does not contain cracks and can be used for metal, concrete and brick surfaces.

4. Conclusions

Thus, SHS spinel-based inorganic pigments are acid- and light-resistant and can be used for color fire-resistant protective and decorative aluminophosphate-bonded coatings.

The coating based on APB and pigment (obtained in the ZnO-MgO-CoO-Al_2O_3 system) consisting mainly of $CoAl_2O_4$ spinel, along with $AlPO_4$ aluminum phosphates crystallized in a hexagonal and orthorhombic crystal systems and $CoAl_2O_4$ spinel, contains acidic aluminum salts with HPO_4^{2-} and $H_2PO_4^-$ anions.

Thermal analysis performed on the SDT Q600 thermal analyzer showed a slight mass loss related to the polymerization of phosphates and the formation of $P_2O_7^{2-}$ anion in the range of 500-550 °C, which is confirmed by IR spectroscopy.

The color of coatings is maintained during annealing at temperatures of 1000 °C.

Metallic particles contained in cobalt oxide - based pigments do not affect the quality of coatings.

References

[1] Maslennikova G N and Pishch I V 2009 *Ceramic pigments* (Moscow: RIF "Stroimaterialy")
[2] Horoshavin L B 2009 *Spinel nano-refractory materials* (Ekaterinburg: UrB RAS)
[3] Ramana Rao M V 2006 *Industrial Minerals* **1** 50-3
[4] Tong Y, Zhang H, Wang S, Chen Z and Bian B 2016 *J. of Nanomaterials* 7 4169673
[5] Lin J, Yu M, Lin C and Liu X 2007. *J. Phys. Chem.* vol 111 **16** 5835-45
[6] Radishevskaya N I, Nazarova A Yu, Lvov O V, Kasatsky N G and Kitler V D 2019 IOP *Conf. Series: Conf. Series PAPER, J. of Phys:* **1214** 012019
[7] Sudakas L G 2008 *Phosphate binding systems* (SPb.: Kvintet)
[8] Gravit M V and Tarabanov V N 2014 *Fire Safety: Problems and Prospects* vol 2 **1(5)** 198-202
[9] Abdrakhimov V Z, Roshchupkina I Yu, Abdrakhimova E S, Kayrakbaev A K and Kolpakov A V 2015 *News of higher educational institutions. Construction* **8** 33-40
[10] RF Patent No.2165948 Pavlov AB and Samoilova AB *Method for obtaining a refractory coating*, publ. 27.04.2001
[11] Ruzinov L P and Gulyanitsky B S 1975 *Equilibrium transformations of metallurgical reactions* (M.: Metallurgy)
[12] Rogachev AS and Mukasyan AS 2013 *Combustion for the synthesis of materials* (M. Fizmatlit) 400p.
[13] Balicheva TG and Lobaneva OA 1983 *Electronic and vibrational spectra of inorganic and coordination compounds* (L. LGU) 120p.
[14] Nakamoto K 1991 *IR spectra and Raman spectra of inorganic and coordination compounds*, trans. from English, ed. by Pentin Yu A (M. Mir) 536p.

IOP Conf. Series: Materials Science and Engineering **665** (2019) 012009 doi:10.1088/1757-899X/665/1/012009

Use of Additive Technologies to Optimize Design of Classifying Devices

A O Romashev

Saint-Petersburg Mining University, 2, 21st Line, St Petersburg 199106, Russia

romashev_ao@pers.spmi.ru

Abstract. The article considers perspectives of the use of additive technologies to optimize the design of classifying devices. A brief overview of the solutions developed using 3D printing of for mining industry is given. It is stated that creation of models is impossible without knowledge of the laws of distribution of flows and expressions describing them. For this purpose, the influence of the pulp viscosity on classification in a hydrocyclone on an experimental unit has been studied. A modified Plitt model was proposed to predict the separation grain. The proposed model takes into account the pulp viscosity and predicts the size of the boundary grain more accurately.

1. Introduction

The rapid development of technology inevitably affects all areas of the modern world. Historically, there are three disrupting common course of "wave" development of the technological revolution, which arose approximately every 50 years since the beginning of the nineteenth century. The era of steam engine is the first wave. The second wave is associated with emergence of mass production in the early 1900s. The third wave took place in 1970s – 1980s, when the industry began to use process control tools, electronics and robotic equipment to automate routine tasks. The fourth wave is built on the basement of the third one, but it is driven by digital technologies, the number thereof is growing exponentially. In particular, the fourth wave is characterized by the use of advanced intelligent systems, robots, neural systems, and additive technologies in production [1-3].

According to experts, the perspectives for the market of additive technologies in Russia are significant. Additive technologies allowed intensifying technologies in many areas of industry, in particular in mining and metallurgical field. Based on the latest equipment, technologies have been developed for production and fine-tuning of high-quality powder materials for additive technologies and powder metallurgy as an element of Industry 4.0 with the use of combined effects - vibration, magnetic and electrical [2-6].

In the field of mineral processing, additive technologies have found their use in optimizing the design of separation devices. For flotation machines, designs of mixing mechanisms have been developed, which mix the pulp, disperse air and create kinetic turbulence. The use of additive technologies improves technological performance, increases flotation flexibility and makes it possible to reduce the rate of rotation of the mixing mechanism, which, in turn, reduces operating costs. In particular, before launching FLSmidth® nextSTEP ™ technology on the market, the company created and tested about 200 rotor/stator combinations using sophisticated software and 3D printing, as well as modeling using computational fluid dynamics (CFD)[3].

Additive technologies play an equally important role in optimizing design parameters of classification in hydrocyclones. For example, the use of small hydrocyclones for separation of particles in microns is of great interest. [4] However, these hydrocyclones are difficult to manufacture because of the need to maintain high precision in manufacturing. Meanwhile, the accuracy and quality of separation largely depends on these parameters, which is confirmed by empirical formulas developed by various researchers. In the case of hydrocycloning as an optimization criterion, it is advisable to use the parameter - boundary grain d_{50c}, which is a complex indicator characterizing operation and performance of the device in general. Using suspensions of pure silica, Plitt [5], Plitt et al. [6] and later Arterburn [7] developed mathematical models linking various factors and d_{50c}. Both of these models were obtained experimentally from experimental data on laboratory-sized hydrocyclones. According to the Plitt model:

$$d_{50c} = \frac{k_1\, D_C^{0,46}\, D_I^{0,6}\, D_O^{1,21} \exp\left(0,063\, C_{VS(F)}\right)}{D_U^{0,71}\, L_{VF}^{0,38}\, Q_{V(F)}^{0,45}\, (\rho_S - \rho_L)^{0,5}}$$

where L_{VF} - free height of the drain pipe (distance from the end of the drain pipe to the sand hole), m; D_C – cylinder diameter, m; D_O, D_I, D_U – diameters of drain, input and sand holes, m; $Q_{V(F)}$ – power volume rate, m^3/sec; $C_{VS(F)}$ - % of solid per volume in feed; d_{50C} - corrected separation size, μm; ρ_S, ρ_L - density of solid and liquid phases, respectively, kg/m^3; k_1 - coefficient depending on the geometric parameters of the hydrocyclone.

Over the years, the Plitt model has undergone many modifications and corrections. Plitt et al. (1980) proposed to add dynamic viscosity of a fluid to the initial model to a power of 0.5, but Austin et al. [8-10] found that the Plitt model gives d_{50c} values that are inconsistent with real data and the model is more suitable for dilute suspensions [11].

In this regard, determination of d_{50c} parameter dependence on the rheological properties of the pulp is an important point for modeling by means of computational fluid dynamics (CFD) and further 3D printing [12-13].

2. Methods

In order to identify dependencies characterizing performance of a hydrocyclone from physical properties of the pulp, a series of experiments were performed in AKW-Laborant ZLF 50ch pilot plant manufactured by AKW Apparate+Verfahren GmbH (Germany). The layout of the installation is shown in Fig.1.

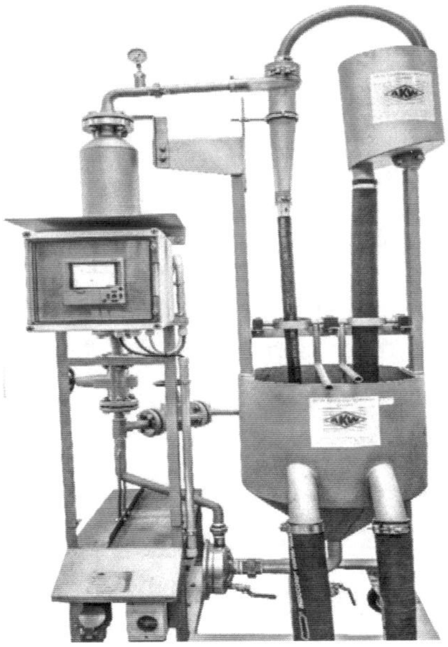

Figure 1. Diagram

The experimental plant comprised a polyurethane hydrocyclone of AKW A+V type, a reservoir for initial power supply, which is tangentially fed into the hydrocyclone using a centrifugal pump (one may install hydrocyclones with a diameter of 80, 55 and 35 mm with variations in the diameter of the sand nozzle), while pulp moves towards hydrocyclone inlet, its density, temperature, and volume flow are measured in a specially installed monitoring device. The pressure at the inlet to the hydrocyclone is measured with a pressure gauge mounted on the feed pipe [9].

Two samples: a mixture consisting of 65% magnetite and 35% quartz, with a particle size of -1+0 mm and a second sample of crushed quartz (80% of class -71 microns) were prepared for experimental studies. The operating mode of the plant was the following: the pressure of the pulp at the inlet to the hydrocyclone is constant (69 kPa); mass fraction of solids in the samples was 19, 26, 35 and 40%; during the study, and viscosity of the pulp was continuously recorded.

3. Results and discussion

The study of the effect of pulp density in the feeding of a hydrocyclone on classification efficiency was performed with a solid content of 10–40% in 5% increments. With an increase in the solid content in the diet, respectively, the density of the pulp was increased. In the course of research, it was found that a higher solid content leads to a decrease in the speeds of the radial motion of solid particles, thus the laws of free centrifugal sedimentation change to the laws of constrained sedimentation. Higher and denser bed of large grains of the solid phase is formed near the sand nozzle. As a result, large grains fall into a drain, and the grain size of d_{50c} boundary grain is increased. There was also a slight increase in hydrocyclone performance with an increase in the pulp density, despite the constant pressure at the inlet to the hydrocyclone, while classification efficiency was lower.

For performance of experiments to establish the influence of temperature and solid content on the pulp viscosity, an experiment involving statistical means will improve accuracy and reliability of the results. The results are shown in Fig.2.

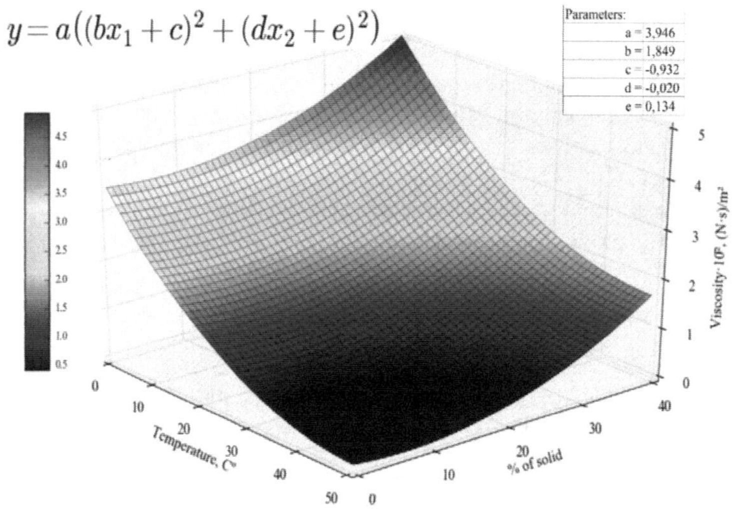

Figure 2. Dependence of the pulp viscosity on the temperature at various concentration of the solid (%)

The obtained functional dependence allows making a correction for the pulp temperature and solid content in the Plitt equation.

In order to assess the effect of viscosity increase, not related to an increase in the solid content in the feed, a series of experiments was performed with a medium viscosity modifier. The concentration of the modifier ranged from 0 to 50% by weight of pulp. The experiments were performed with a constant solid content in the feed of 10% and temperature with a small range of variation: 20–23°C.

In the course of the experiment, the effect of the pulp viscosity on the size of d_{50c} grain was estimated. The results of the study are shown in Fig.3.

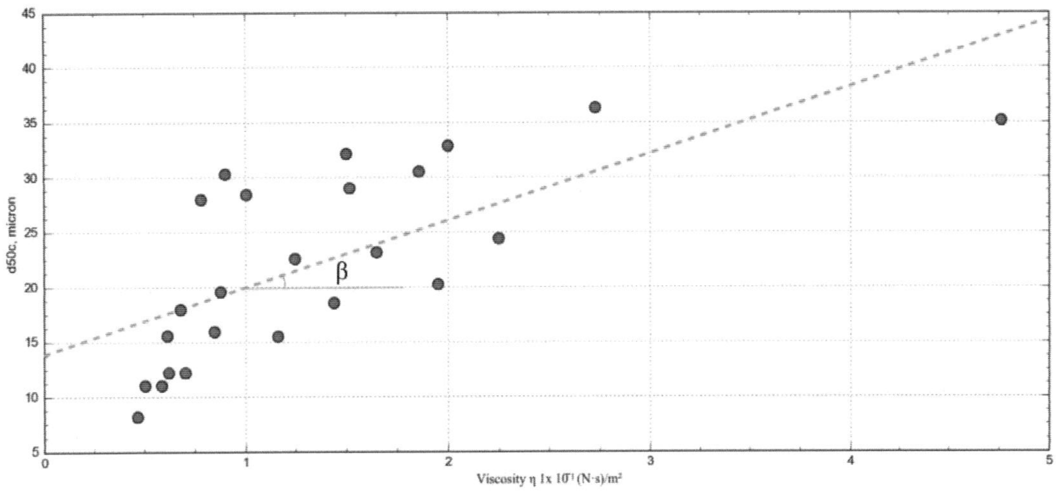

Figure 3. Dependence of pulp viscosity on d_{50c}.

The dependence of the grain size of the boundary grain on the viscosity of the pulp obtained in the course of experimental studies was subsequently introduced into the Plitt model described above. A new model was formulated, first, by introducing the term of viscosity and then changing the term of solid percentage in the Plitt model [5]. In the original Plitt model, the term is used as a different percentage of solid

The original Plitt model shows this term as ($0,063C_{VS(F)}$), and it was replaced with $C_{VS(F)}{}^{0,41}$ in the new equation:

$$d_{50c} = \frac{k_1\, D_C^{0,46}\, D_I^{0,6}\, D_O^{1,21}\, C_{VS(F)}{}^{0,41} \eta^{\beta}}{D_U^{0,71}\, L_{VF}^{0,38}\, Q_{V(F)}^{0,45}\, (\rho_S - \rho_L)^{0,5}}$$

where: k_2 – coefficient depending on rheological properties of the pulp and geometric parameters of the hydrocyclone; η - pulp viscosity, mPa sec; β – degree of viscosity effect.

Indicator β can be determined graphically and is numerically equal to the slope of the regression prima.

As noted above, the Plitt model is designed to measure the pulp viscosity only from a percentage of solid, and is not designed for any changes in viscosity from other sources (temperature, for example). In fig. 4 shows, how changes in d_{50c} from both temperature and solid% are accurately predicted using a new model.

This equation uses viscosity data obtained by direct measurement, an increase in d_{50c} due to changes in viscosity, both from an increase in the solid content (for different % of solids) and temperature changes (from 10°C to 50°C in the individual range of % solids)).

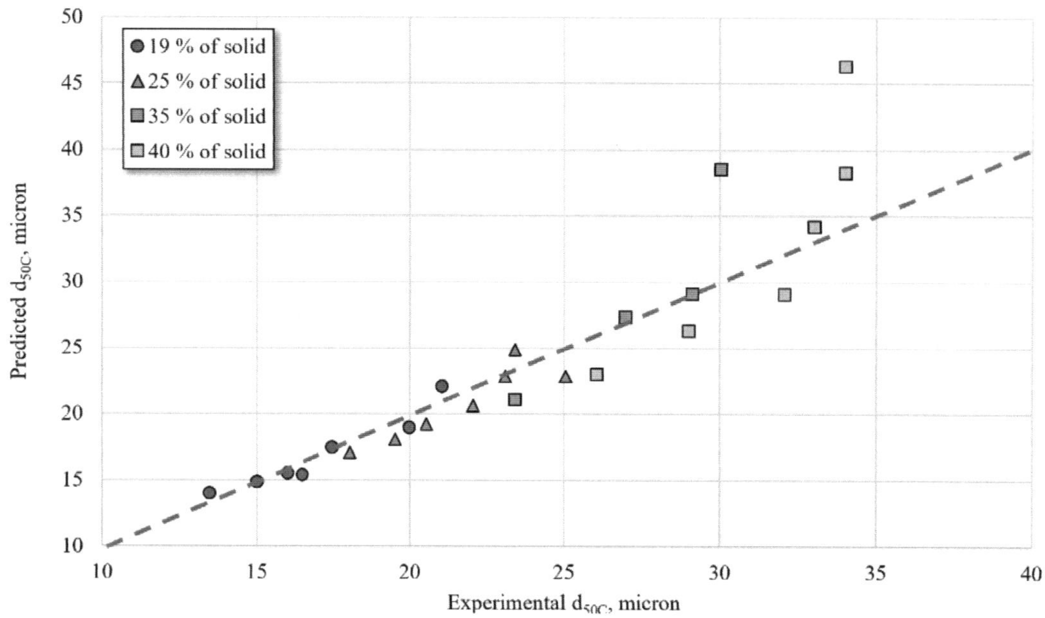

Figure 4. Predicted value d_{50c} for the new modified equation as a function of d_{50c} being observed.

4. Conclusion

Viscosity plays a major role in classification in a hydrocyclone. The viscosity of the pulp may vary under the influence of several factors, e.g.: any changes in the solids content, particle size, temperature, chemical environment, etc. Therefore, the viscosity index should be considered in addition to the solids content in the pulp, when creating models for the hydrocyclone separation grain.

Thus, in the course of experimental studies, dependences were obtained describing the effect of rheological properties on its performance as a classifier. The coefficient that corrected the viscosity in the Plitt model, which allows increasing the accuracy of the calculation, has been determined experimentally. The results are an excellent prerequisite for creating high-precision CFD models. The creation of such models will allow optimizing the design parameters of the hydrocyclone followed by production of the whole structure or its part by means of 3D printing.

Acknowledgments

This work was supported by a grant from the President of the Russian Federation, MK MK-5915.2018.5.

References

[1] Chanturiya V A , Vaysberg L A and Kozlov A P 2014 *Obogashchenie Rud,* **2** 3-9.
[2] Andronov G P, Zakharova I B, Filimonova N M, Lvov V V and Aleksandrova T N 2017 *Journal of Mining Science* **52-6** 1190-4.
[3] Talovina I V, Lieberwirth H, Alexandrova T N and Heide G 2017 *Eurasian Mining,* **1** 21-4.
[4] Vaisberg L, Ustinov I 2017 *Geoinformatics Research Papers* **5-1** 28.
[5] Aleksandrova T N, Heide G and Afanasova A V 2019 Journal of Mining Institute 235 30–7.
[6] Afanasova A, Aleksandrova T and Nikolaeva N 2018 SGEM, 18 (1.4) 119-24.
[3] Leonida C 2015 *Mining Magazine,* 1261556
[4] Vega-Garcia D Brito-Parada P R and Cilliers J J 2018 *Chemical Engineering Journal.* **350** 653-9
[5] Plitt L R 1976 *CIM Bull.* **69** 114.
[6] Plitt L R , Finch J A and Flintoff B C 1980 *Euro. symposium particle tech.* 790–804.

[7] Arterburn R A 1982 *Design and installation of communication circuits*. ed: Mular A L, Jorgensen G V (New York: AIME)592–607.
[8] Austin L G, Klimpel R R and Luckie P T 1984 *Process engineering of size reduction: ball milling*
[9] Lvov V V and Aleksandrova T N 2016 *Gornyi Zhurnal [Mining Journal]*, 5 94-6.
[10] Nikolaeva N V, Aleksandrova T N and Elbendari A M 2018 *Geomechanics and Geodynamics of Rock Masses*, **1** 333-8.
[11] Gupta A, Eren H 1990 *Proce AusIMM*, **295(2)** 31.
[12] Aleksandrova T N, Prokhorov K V and Lvov V V 2015 *Gornyi Zhurnal*, **12** 4-8
[13] Korchevenkov S A, Aleksandrova T N 2017 *Metallurgist*, **61 (5-6)** 375-81.

IOP Conf. Series: Materials Science and Engineering **665** (2019) 012010 doi:10.1088/1757-899X/665/1/012010

Simulation and analysis of the shaft strength of a centrifugal pump in an in-situ leaching method of uranium

M P Sadykov

L.N. Gumilyov Eurasian National University, System Analysis and Control Department, 11 Pushkin Street, Nur-Sultan (Astana), 010000, Kazakhstan
Email: sadykov.mahambetp@gmail.com

Abstract. This research shows the performance of the strength analysis and dynamic analysis of the shaft strength of a centrifugal pump in an integrated NASTRAN / PATRAN system. There is conducted the calculation of a mechanical scheme for determining the stress-strain state of the rotor shaft of a centrifugal pump and constructed a 3D-model of the rotor of a seven-step centrifugal pump. Furthermore, there are determined voltage levels for zero and maximum flow. On the basis of these dynamic calculations, a damping element has been proposed that ensures a smooth entrance of CP into an operating mode - a magnetic coupling.

1 Introduction

An in-situ leaching technique (ISL), usually used in the extraction of uranium. A solution of the sulfuric acid of low concentration, or another oxidizing agent (for example, formation water saturated with oxygen in the air) is pumped through an injection well into a formation and filtered through a uranium-containing rock, while gradually dissolving it. Then, a metal-enriched solution is pumped to a surface through a pumping well, leaving mostly an intact underground mountain system, i.e. the metal mining is carried out at the place of the occurrence of ores. Subsequently, uranium is extracted on special technological equipment (sorption on ion-exchange resin), and a leach solution is recycled. Before re-supplying the leach solution to a formation, it is enriched with leaching reagents.

In-situ leaching accounts for about 20% of the world uranium industry. A leadership yet retain underground mines (40%) and open careers (30%). However, the fact says that such advanced uranium mining countries as the United States, Uzbekistan and the undisputed leader of the industry, Kazakhstan, give preference to ISL [1].

Pumps are one of the key elements of an in-situ leaching technology: they are involved in all stages of production. That is why its quality and durability take the first place in the selection of equipment - breaks in work are unacceptable, as well as too few hours of work. Indeed, in the latter case, the pumps have to be changed frequently, and these are additional financial and time costs [2].

As part of the research work, there has been created an optimized design of a central pump on the basis of the model of a pump brand 'ODDESSE zentralasien' - UPP 13-7 / 6. The task of the work is to increase the efficiency of the central pump from an existing value of 40% to a reference value of 60%. To achieve this goal, one of the first stages of research is to conduct an automated strength calculation in PATRAN.

In the practice of pump design, centrifugal sizes are most often set not from a strength condition, but from design considerations and depending on technological capabilities. In addition, new pumps are usually designed on the basis of tested prototypes. Therefore, when designing, the main importance is the verification calculation of the strength of the main elements of a pump: shaft, impeller, housing, keyway or pin joint, coupling [3].

The task of the verification calculation is to determine the values of parameters with which in each case the loss of the strength (destruction) of a particular element (normal or tangential stresses,

IOP Conf. Series: Materials Science and Engineering **665** (2019) 012010 doi:10.1088/1757-899X/665/1/012010

deformations, rotational speed, etc.) is associated with their subsequent comparison with some limit permissible values.

The automated strength analysis of the rotor shaft of a pump has been carried out in a NASTRAN / PATRAN program.

For a calculation, there has been simulated a calculated mechanical scheme according to a method:

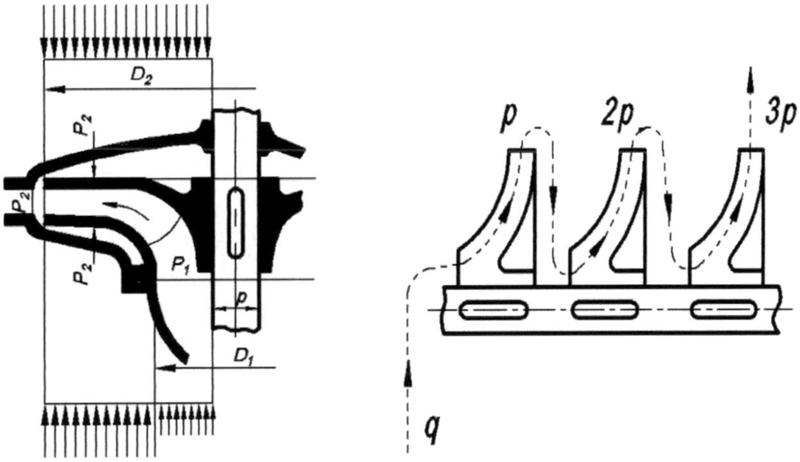

Figure 1. Design diagram a) pump stages b) pressure distribution in pump stages

Axial and radial forces act on a pump stage. The axial force acting on the impeller of a pump is a resultant force acting on an impeller and it is directed in a direction corresponding to the direction of the action of the larger of these forces.

2. Calculation of forces acting on impellers

Testing for the static strength of a shaft is carried out on the greatest loads that may occur in the various modes of pump operation - this is a mode with zero flow (or close to zero) and a mode with a maximum flow allowed during operation. The fixed scheme of a vertical submersible pump is shown in Figure 2.

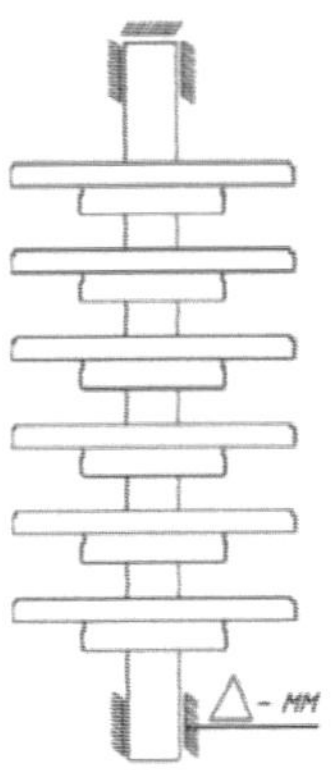

Figure 2. Circuit diagram of a centrifugal pump

The lower end of a shaft is worn in a nylon sleeve and it is free until the start of a suction. Later, when modeling in PATRAN, this boundary condition has been modeled as a degree of freedom allowing the movement of Δ according to an assembly's working drawing Δ = 4 mm.

3. Definition of axial force

Axial force can be approximately defined as the difference in pressure forces to right and left from R_l to R_b [4].

$$P_{axial} = \frac{\pi}{4\left(R_1^2 - R_b^2\right)H} \cdot Y \quad (1)$$

where, P_{axial} – axial force in H;

R_l – outer radius of the inlet of an impeller m;

R_b – shaft radius in m;

H – pump head in m;

Y – volume of liquid in kg / m³.

IOP Conf. Series: Materials Science and Engineering **665** (2019) 012010 doi:10.1088/1757-899X/665/1/012010

Considering that a centrifugal pump is designed for pumping sulfuric acid with a density of 1050 kg / m^3, there has been found the value of axial force.

4. Definition of radial force

The main cause of shaft deflection is a radial force. For centrifugal pumps without a spiral outlet and when connecting sections with ties, the radial force arises from the dynamic unbalance of a rotor.

The dynamic imbalance of a rotor is a consequence of the inaccuracy of the manufacturing parts of the rotor (primarily with large radial dimensions, such as impellers). The presence of an imbalance of rotor parts during the rotation of a rotor leads to the appearance of dynamic loads.

It is possible to determine a strength at a known residual unbalance of an impeller by the following formula:

$$R_D = me\omega^2 \quad (2)$$

where, mR_2 – permissible sufficient imbalance of an impeller, kg · m (for example, an expression 'Permissible residual imbalance of 100 g · mm' means that for the impeller with a radius of 100 mm, balancing on an external radius should be carried out with an accuracy of 1 g, which is technologically achievable);

ω – angular velocity of the rotation of a rotor pump, 1/s [5].

Table 1. Eccentricity with different impeller diameters.

Impeller outer diameter, mm	< 300	300-500	500-1000	1000-2000
Eccentricity, mm	0.075	0.100	0.150	0.200

To calculate a radial force from imbalance, it is used SolidWorks capabilities to determine the mass of a 3D model, in our case for seven sections, m = 5.43 kg.

Considering the principle of the superposition of force, the radial force has been calculated as the sum of force from imbalance at each stage of a rotor.

Taking into account the pressure distribution scheme along with the steps of a centrifugal pump Figure 3, and using the above method of calculating radial and axial loads, there have been determined boundary conditions, input data for modeling the strength of a rotor shaft and impellers are given below:

Table 2. Input simulation data the strength of a rotor shaft and impellers.

No	Parameter	Value
1	Material	Steel
2	Modulus of elasticity E	$2 \cdot 10^{11}$ PA
3	Poisson's ratio	0,3
4	Support 1	Hinge with rotation around the axis of a model
5	Support 2	Hinge with rotation around the axis of a

IOP Conf. Series: Materials Science and Engineering **665** (2019) 012010 doi:10.1088/1757-899X/665/1/012010

		model
6	$P_{1axial} = 232$ H $P_{2axial} = 464$ H $P_{3axial} = 696$ H $P_{4axial} = 928$ H $P_{5axial} = 1160$ H $P_{+6axial} = 1392$ H $P_{7axial} = 1624$ H	
7	Radial force	744 H
8	Type of calculation	Static

5. Creating a model in PATRAN

A geometric model has been created directly in PATRAN, there has been generated from TET elements, force has been applied in radial and axial direction through a 'Force' command to a node, boundary conditions for fixings have been created with conditions Displacement <0,0,0>, Rotation <0, ,0> for upper and lower support, and for zero feed Displacement <0,0,0>, Rotation <0, ,0> for upper support, Displacement <0,0,0>, Rotation <0,1.4,0> for lower support.

Figure 3. Finite element model of a pump shaft

6. Results analysis

The results of the calculation of equivalent stresses are shown in Figure 4.

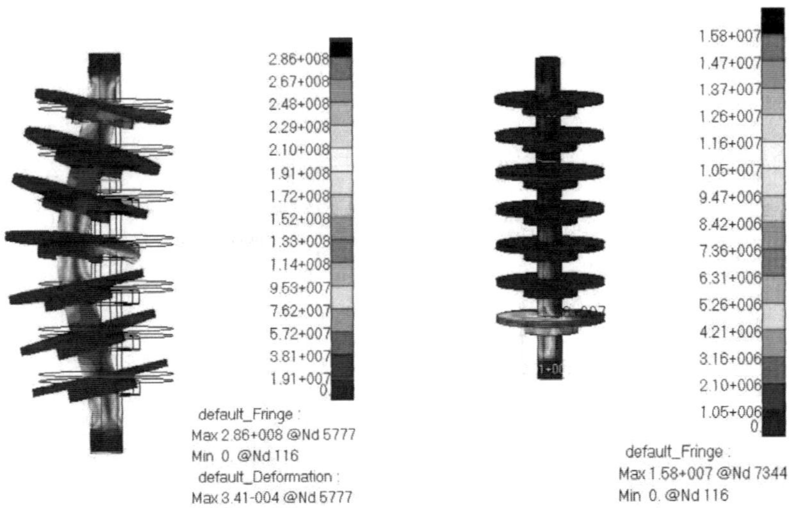

Figure 4. Voltage map for zero and maximum flow

An analysis of the results showed a sufficient safety factor at maximum flow, according to the diagram σmax = 286 MPa, the allowable stress for steel 40X $[\sigma -]$ = 680 MPa, a safety factor was 2.37.

The results of displacements at zero flow showed the adequacy of a computer scheme created in PATRAN, $y_{max} = 4 - 10^{-3}$, which corresponds to boundary conditions.

Figure 5. Shaft motion chart.

The results of a strength analysis showed that the shaft of a newly designed centrifugal pump can be made of a less expensive material than 40X steel.

Increasing the pressure-energy performance of machines for pumping a liquid media is impossible without optimizing the dynamic parameters of a system at a design stage.

The reliability and service life of a centrifugal pump is largely determined by its vibration state. A technology for calculating the critical rotor speed of a pump is complex, and so far there is no possibility to accurately determine it due to the impossibility of the reliable prediction of coefficients that take into account an effect of all possible factors affecting the vibration state of a pump [4].

7. Theoretical provisions of a numerical analysis of rotor dynamics

Free vibrations completely determine the dynamic properties of a mechanical system and have paramount importance in analyzing forced vibrations [5], therefore using a finite-element model, firstly, they determine the spectrum of the natural frequencies of the rotor of a centrifugal pump.

To describe a movement only under the action of a restoring (elastic) force without, taking into account energy dissipation, the equation is used [5]:

$$[M]\{\ddot{q}\}+[C]\{q\}=0 \quad (3)$$

where, [M], [C] – matrix of masses (inertia) and the rigidity of a system; $\{\ddot{q}\},\{q\}$ – generalized displacements of nodes and their derivatives.

The solution of equation (3) is sought in the form [7]:

$$\{q\}=\{q_0\}\sin\omega_0 t(2\text{-}17) \quad (4)$$

where, ω_0 – values of the natural frequencies of oscillations;

$\{q\}$ – full vector of the nodal displacements of a system;

$\{q_0\}$ – amplitude column matrix.

A full vector q is a function of independent displacement components and rotation angles about corresponding axes. The full displacement vector is represented as:

$$\{q\}=\left\{\{q^{(1)}\}\cdot\{q^{(2)}\}...\{q^{(n)}\}\right\}^T \quad (5)$$

In this case, a task is reduced to calculating the eigenvalues of frequencies ω_0 and the eigenvalues of the vectors of a generalized displacement q, hence, q determines the shape of natural oscillations at a corresponding frequency ω_0. When implementing an automated finite element method for determining natural oscillations, a numerical solution of the system of algebraic equations in the algorithms of a PATRAN program [91,92,103,104] is carried out using a Lanczos method [6].
The forced oscillations of a rotor occur under an action of the harmonic centrifugal force of the inertia of the unbalanced masses of a rotor, which are represented as $F_u = m\omega^2 e\cos(\omega t)$, then the equation of forced oscillations is written:

$$[M]\cdot\{\ddot{q}\}+[B]\cdot\{\dot{q}\}+[C]\cdot\{q\}=[Me]\omega^2\cos(\omega t) \tag{6}$$

where [M], [B], [C] are a matrix of masses (inertia), damping and the rigidity of a system;
$\{q\}$, $\{\ddot{q}\}$, $\{\dot{q}\}$ – generalized displacements of nodes and their derivatives;
ω – angular velocity of a rotation;
e – specific imbalance.
The solution of an equation (6) is sought as:

$$\{q\}=\{q_0\}\sin\omega_0 t + [Me]\,\omega^2\cos(\omega t) \tag{7}$$

8. Automated calculation of the natural frequencies and oscillations of the rotor of a centrifugal pump

Initial data for a calculation are the physical properties of a shaft material (density ρ = 7850 kg / m³ and the modulus of the elasticity of the 1-st kind E = 2.1 · 1011 N / m²), lengths L, outer D and inner d diameters of sections, mass m, and also stiffness from bearing support.
According to the design scheme (Figure 6), a beam model has been built in Patran, which includes 7 CBEAM elements (rotor shaft), 7 CONM2 elements (concentrated mass element simulating a rotor wheel).
The modeling of rigid supports is carried out by fixing rotor model nodes according to the corresponding degrees of freedom; in the design of a pump under study, the support was modeled taking into account a gap Δ - mm (Figure 6).

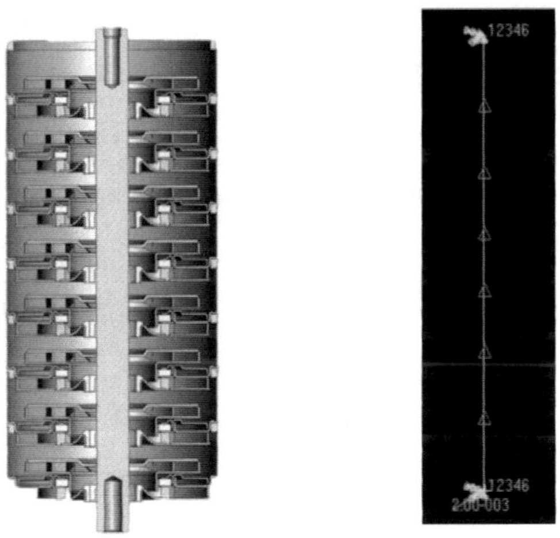

Figure 6. Centrifugal pump rotor, the rod model of a rotor, built in Patran

The values of the natural frequencies of a rotor have been found using the NORMAL MODELS solver (modal analysis).

Table 3. Found values of natural frequencies.

Parameter	Value

f_1 - first natural frequency Hz	8
F_2 - second natural frequency Hz	434
F_3 - third natural frequency Hz	447
F_4 - third natural frequency Hz	696
F_5 - third natural frequency Hz	732
F_6 - third natural frequency Hz	892

The coincidence of often disturbing oscillations with the frequencies of natural oscillations presented in a table and can lead to resonant phenomena.

9. Automated calculation of perturbed oscillation frequencies

To determine the frequency of disturbing oscillations, there has been used a COMPLEX EIGENVALUE solver (complex frequencies). A program can use the option of asynchronous precession (ASYNC) to determine the response of a system to external influences, which is independent of rotor speed. When using synchronous precession (SYNC) option, the system's response to an imbalance or other excitation is determined, which depends on the rotor speed. With the help of the complex analysis of forms, it is possible to determine oscillation frequencies corresponding to direct and inverse precession, as well as critical rotational speed.

In the Spin Profile menu, a user sets individual rotor speed; for our rotor task of a centrifugal pump, an angular velocity ω = 3000 rpm. It also takes the moments of the inertia of mounted elements that have been defined in a SolidWorks CAD system.

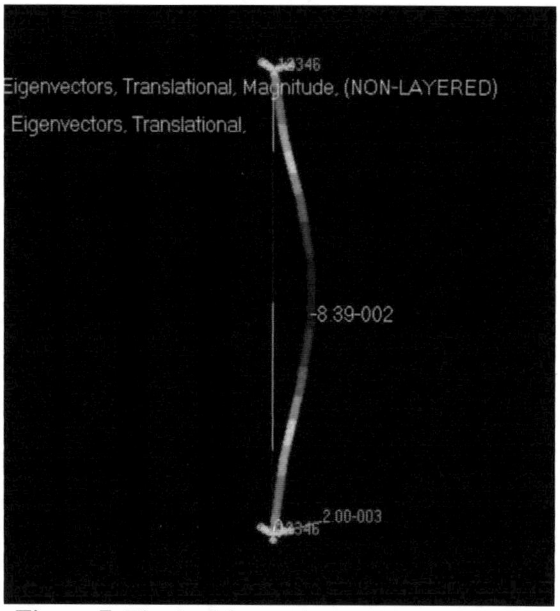

Figure 7. The 3-rd form of perturbed oscillations at a frequency of 1800 rpm.

When choosing the type of calculation, the calculation of the complex eigenvalues of SOL 107 is a direct method, the frequency diagram is obtained by calculating complex eigenvalues by the direct method using an ASYNC option, at the rotor speed of 0, 900, 1800, 2700, 3000 rpm [7].

Table 4. Found values of perturbed oscillations.

Parameter	Value
f_1 - first frequency Hz	0
F_2 - second frequency Hz	463
F_3 - third frequency Hz	488

F_4 - third frequency Hz	732
F_5 - third frequency Hz	736

Critical speed is determined based on which eigenvalues are identical to the speed of the rotation of a rotor. For this purpose, a straight line is constructed on a diagram, corresponding to w = W, i.e. (oscillation frequency = angular speed of the rotation of a rotor). The points of the intersection of a line with the curves of natural frequencies correspond to critical rotor speed [10].

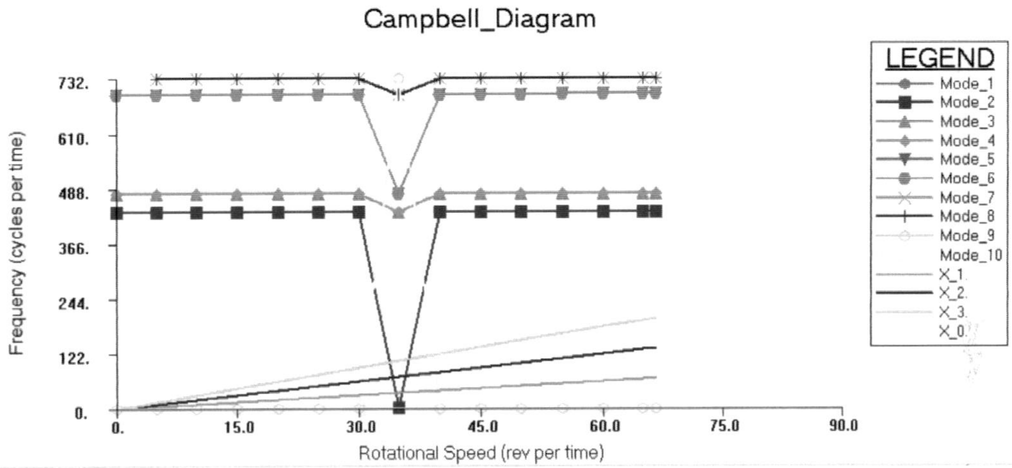

Figure 8. Campbell chart.

The broken character of frequency lines indicates the unstable nature of vibrations associated with the design features of guide supports, in the form of a gap of a size Δ.

Automatically calculated eigenvalues (Figure 8), corresponding to the same oscillation forms, form a series of curves that are the functions of changing an oscillation frequency from the angular velocity of the rotation of a rotor. A Campbell diagram shows that all repeated critical rotational frequencies for the first modes of oscillation – 366, 488, 732 Hz are in the region of 47 Hz (2200 rpm) for reverse and direct precession, respectively. To compensate for vibrations during the start-up period, it is planned to replace a finger clutch with a magnetic clutch with better damping properties.

An operating mode has been modeled by a design scheme with support imposing restrictions on displacements in a plane perpendicular to the axis of a rotor shaft axis in reality – this layer of distilled liquid filling a pump cavity. The task has been calculated with the same input data as the circuit presented above.

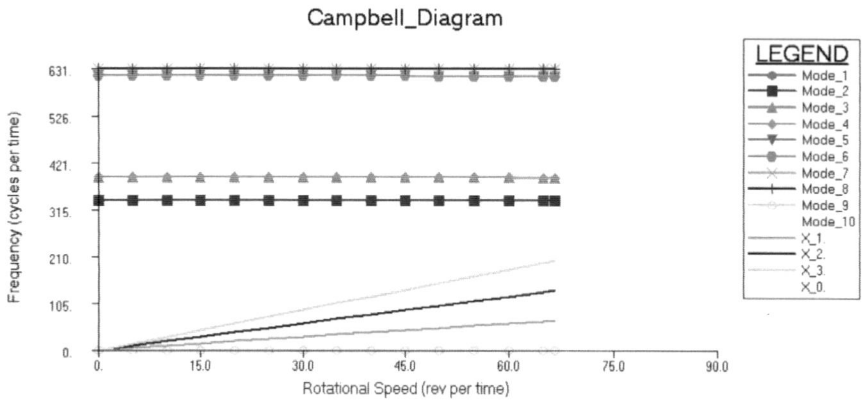

Figure 9. Campbell diagram for a model with support without a gap.

Table 5. Frequency of disturbed oscillations in the second case of consolidation is summarized in a table

Parameter	Value
f_1 - first frequency Hz	0
f_2 - second frequency Hz	315
f_3 - third frequency Hz	388
f_4 - third frequency Hz	631
f_5 - third frequency Hz	636

Despite the fact that the tones of the oscillations of the second model are 20% lower than those in guide bearings, it can be said that critical speed points (points of the intersection of speed multiplicity lines and frequency lines) are missing and a machine enters an operating mode relatively stable.

10. Conclusion

In an integrated NASTRAN / PATRAN system, strength and dynamic analysis have been performed and the following results have been obtained:

- calculated mechanical schemes for determining the stress-deformable state of the rotor shaft of a centrifugal pump;
- 3D model of a seven-stage centrifugal pump rotor has been built;
- there have been determined voltage levels for zero and maximum flow σmax = 286 MPA for maximum flow, σmax = 158 MPA for minimum flow.

The dynamic parameters of a rotor shaft have been analyzed in two operating modes in a COMPLEX EIGENVALUE solver of a ROTOR DYNAMICS NASTRAN module, the determination of the frequency of disturbing oscillations, critical rotational frequencies for the first oscillation forms – 366, 448, 732 Hz are in the region of 47 Hz (2200 rpm) for reverse and direct precession, respectively. On the basis of the dynamic calculations, a damping element has been proposed that ensures a smooth entrance of CP into an operating mode - a magnetic coupling.

References

[1] Li L, Sagüés A A and Poor N 1999 *J. Cement and Concrete Research* **29** 315–21
[2] Haque N and Norgate T 2014 *J. of Cleaner Production* **84** 382–90
[3] Mudd G M 2001 *J. Environmental Geology* **41** 390–403

IOP Conf. Series: Materials Science and Engineering **665** (2019) 012010 doi:10.1088/1757-899X/665/1/012010

[4] Seredkin M, Zabolotsky A and Jeffress G 2016 *J. Ore Geology Reviews* **79** 500–14
[5] Uhrie J L, Drever J I, Colberg P J S and Nesbitt C C 1996 *J. Hydrometallurgy* **43** 231–39
[6] Seaman J C, Bertsch P M and Schwallie L 1999 *J. Environ. Sci. Technol.* **33** 938–44
[7] Miró M, Hansen E H, Chomchoei R and Frenzel W *J. TrAC Trends in Analytical Chemistry* **24** 759–71

IOP Conf. Series: Materials Science and Engineering **665** (2019) 012011 doi:10.1088/1757-899X/665/1/012011

Analytical description of stress-strain dependence for segmented polyurethane ureas swollen in physically aggressive media

V Yu Senichev, M A Makarova, E R Volkova, A V Savchuk and A I Slobodinyuk

Institute of Technical Chemistry, a branch of the Perm Federal Research Center of the Ural Branch of the Russian Academy of Sciences, 3 Akademika Koroleva Street, 614013, Russia

E-mail: senichev85@yandex.ru

Abstract. The article is devoted to the problem of analytical description of the strain behavior of elastomers under various conditions. It summarizes experimental data on the stress versus strain dependences of polyurethane urea elastomers swollen in two plasticizers. These data can be adequately described using the approach taking into account relaxation properties of elastomers and non-relaxation ones. All factors affecting the strain behavior of investigated elastomers are considered.

1. Introduction

Elastomers based on polyurethane ureas and polyurethanes are widely used in industry due to the advantages of their physic-mechanical properties comparing with usual rubbers. The most complete information about the deformation behavior of these materials can be obtained using the experimental dependence of stress versus strain. Such dependences can be obtained as analytical equations using the previously proposed generalized approach to the theory of elasticity [1, 2].

However, when the mentioned materials are used in physically aggressive media (solvents, plasticizers), their properties significantly vary. It was previously established, for example, that the dependence of the initial modulus of the swollen amorphous cross-linked elastomers on the plasticizer concentration can be described by a complex function. Herewith the parts of the modulus related with chemical bonds and physical ones depend on the plasticizer concentration according a power function with exponent m [3]. This exponent is equal to 1/3 for chemical bonds, but $m>2$ for physical bonds correspondingly. The value of this exponent is related to the nature of the plasticizer, the polar groups of which are able to interact with the polar groups of the polymer, competing with the formation of inter-chain physical bonds in the polymer system.

It should be noted that the most important polyurethanes and polyurethane ureas are block-copolymers, i.e. polymers whose polymer chains consist of alternating segments that differ greatly in chemical structure. This difference leads to the phase separation, promoting formation of hard and soft block domains [4]. The domains of hard blocks in such elastomers play, on the one hand, the role of reinforcing filler, and on the other hand, they play the role of spatial network nodes. Under the conditions of swelling hard domains may be partially or completely break down [5].

The target of this work is investigation of the possibility of analytically describing the dependence of stress versus strain for segmented polyurethane urea elastomers swollen in plasticizers.

2. Materials and Methods

We investigated the polyether urethane urea elastomers with different content of hard segments. All of them were made on the basis of oligotetramethylene oxide diol (PTMO-1400) with M_n = 1400 g·mol^{-1} and 2,4-toluene diisocyanate (TDI). Methylene-ortho-chloroaniline (MOCA) was used as a chain extender, and trimethylolpropane (TMP) as a cross-linking agent. Di-2,6-hexyl sebacate (DEHS) and tributyl phosphate (TBP) were used as plasticizers.

Synthesis of elastomers was carried out using two stages. At the first stage of synthesis, a prepolymer was obtained by the reaction between PTMO-1400 and TDI. In this case, the ratio NCO/OH = 2.06 was applied. In the second stage, prepolymers were reacted with mixtures of the aromatic diamine MOCA, PTMO-1400 and TMP. The ratio of components in the synthesis are shown in Table 1. Preparation of the samples was published earlier [6].

Table 1. Molar ratios between components of compositions.

Series	NCO/OH	PTMO-1400	TMP	MOCA
PT-1	1.08	0.95	0.05	0
PT-2	1.08	0.75	0.05	0.2
PT-3	1.08	0.55	0.05	0.4
PT-4	1.08	0.15	0.05	0.8

Cured elastomer samples were cut to obtain rectangular plates for further swelling to achieve a certain swelling degree. The amount of this degree for each sample in the DEHS was chosen so that the values of the polymer volume fraction in the swollen samples φ_2 were distributed approximately uniformly in the range of 1÷0.66. Subsequently, the swelling in the TBP was carried out to achieve the values of φ_2, which were almost equal to the ones swollen in DEHS.

Mechanical testing of elastomers was performed using Instron 3365 testing machine. The tests were carried out at 25±1 ° C under three stretching rates 0.0028 s^{-1}, 0.056 s^{-1}, and 0.56 s^{-1}. The relaxation characteristics of the studied samples were determined by the Bartenev-Bryukhanov method [7] under the same stretching ratio $\lambda = 1.4$.

3. Results and discussion

As mentioned earlier, the stress versus strain dependence for the cross-linked elastomers can be described using the previously developed approach that takes into account both the relaxation properties and non-relaxation ones of the elastomers. This approach gives the following equation for stretching at a constant rate [2]:

$$\sigma = E_0(\lambda - \lambda^{-2}) \cdot \left[1 + 3b(\lambda^{-1} - d) + \gamma_s \alpha^2 (1-\phi)(\alpha^2\phi - 2)/(1-\alpha^2\phi)^2\right] + \\ + \int_1^{\lambda} W \sum_{i=1}^{n} E_i \exp\frac{1-\lambda}{v\tau_i} d\lambda = \sigma_{el} + \int_1^{\lambda} W \sum_{i=1}^{n} E_i \exp\frac{1-\lambda}{v\tau_i} d\lambda \quad (1)$$

where σ_{el} is the highly elastic part of the stress and not related to the relaxation properties of the object, λ is the stretching ratio, v is the stretching rate, τ_i is the relaxation time, E_i is the relaxation parameter with stress dimensionality, W is the transformation function, α is the chain inextensibility coefficient, and ϕ is a intermediate function, $\phi = \lambda^2 + 2\lambda^{-1}$. E_0 can be considered as a parameter of a relaxation element with an infinite relaxation time and identified as the elastic response of a constant spatial network with a certain density. The change in the freedom degree of polymer segments in the deformation process of polymer chains is taken into account by introducing α, and the γ_s coefficient reflecting the fraction of polymer chains taking part in the deformation process in an overstressed state.

Parameters *b*, *d* are chemical network constants for specific series. Usually $d \approx 0.2$ for unswollen elastomers.

When adapting equation (1) to the case of swollen polymer networks, it is possible to use semi-empirical dependencies of some parameters on the swelling degree. Thus, these dependences were obtained in [8] for the *d* and E_i quantities. In principle, the question of the swelling effect on the mentioned parameters remains open. However, the approximating analysis of the swelling effect on these parameters can be done using the Bueche approach which relates the strength and tensile properties of polymer networks with the rupture probability of overstressed polymer chains [9].

Let us consider, for example, in the simplest form, swelling of three-dimensional networks within the framework of the Flory's theory, setting as a constant criterion related to the magnitude of the limiting strain degree, the specific work of strain *A* performed upon reaching the $\lambda_{\max}$ value:

$$A = \lambda_{\max}^2 + 2\lambda_{\max}^{-1} \tag{2}$$

Taking into account the fact that a part of the specific work of strain during deformation of swollen networks spend for three-dimensional stretching (during swelling), we can obtain the following formula connecting the values of the maximum stretching ration before swelling and after swelling (λ_{0max} and λ_{max}):

$$\lambda_{0\max}^2 + 2\lambda_{0\max}^{-1} = \lambda_{\max}^2 + 2\lambda_{\max}^{-1} + 3(1 - \varphi_2^{-2/3}) \tag{3}$$

Calculations using this equation with transition to α values ($\alpha = 1/\lambda_{max}$), show that the variation in α values during swelling can be neglected within the diapason of small swelling degrees (up to 30%) and for relatively low ultimate stretching values ($\lambda_{0max} < 7$). Since the γ_s parameter plays a subordinate role with respect to α, one can also assume its permanent status under swelling in a similar way.

On the other hand, it is obvious that plasticization can affect all the relaxation parameters included in this equation. In this work, we used the approach described in [8], in which the values of the characteristic relaxation times of the polymers under consideration are frozen, which to some extent corresponds to the principle of continuity of the relaxation time spectrum of most polymers. In this case, the swelling effect can be marked only for the parameters of the corresponding relaxation elements having the modulus dimensionality. Let us now consider the dependence of the above mentioned parameters and all the stress versus strain plots on the swelling degree, i.e. on the φ_2 values.

Table 1 shows the relaxation parameters of the studied elastomers and the features of their variation in the swelling process. These variations for the E_i values were considered in the framework of the following dependencies:

$$E_i = E_{i0}\varphi_2^m \tag{4}$$

Generally, a large volume of experimental data was obtained, which made it possible to draw a number of important conclusions. Firstly, the used equation (1), taking into account the concentration correction of the E_i parameters, makes it possible to obtain calculated stress versus strain plots. Parameters of the samples necessary for calculations are given in Tables 2, 3. They were determined according to [1]. As shown in Fig.1, the agreement between the experimental plots and calculated ones is quite satisfactory, which allows us to conclude that the corresponding approach is unconditionally applicable for swollen segmented elastomers.

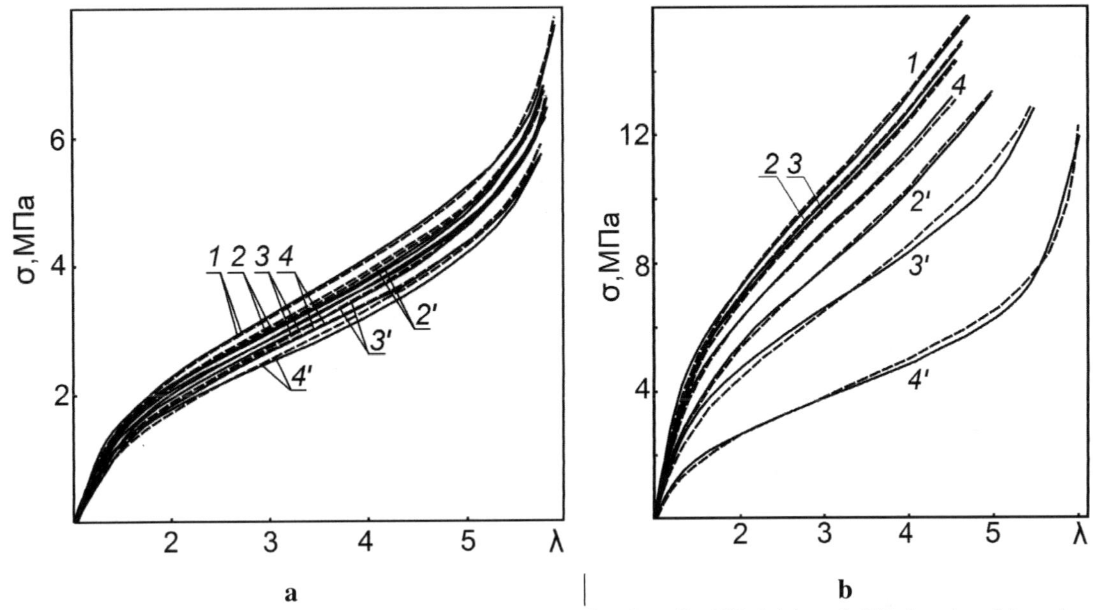

Figure 1. Comparison of calculated stress versus strain plots for PT-1 (a) and PT-4 series (b) under 0.56 s^{-1} stretching rate. Solid curves are experimental, dotted ones are calculated using equation (1)

Table 2. Relaxation parameters of the unswollen PT elastomers at 25°C

	Series	E_0, MPa	E_1, MPa	E_2, MPa	E_3, MPa	τ_1, s·10^{-3}	τ_1, s·10^{-2}	τ_1, s
	PT-1	0.84	0.18	0.14	0.17	1.02	1.10	10.2
	PT-2	0.88	0.18	0.14	0.17	0.99	1.12	9.9
	PT-3	0.93	0.18	0.14	0.17	1.01	1.11	9.5
	PT-4	2.93	0.18	0.14	0.17	1.02	1.10	9.9
	DEHS	0.33	2.2	2.2	2.2			
m	TBP	2.3 (0.33 for PT-1)	3.5	3.5	3.5			

Also some important conclusion can be made considering values of exponents *m* for the E_i parameters presented in Table 1. The average value of the index *m* for the E_1-E_4 parameters for samples swollen in DEHS is 2.2, and the same for ones swollen in TBP is 3.5. Similar phenomena for relaxation elements caused by physical bonds have already been described previously [8].

But we can consider now the plasticization effect on the network of strong physical bonds caused by the presence of the hard domains in polyurethane ureas. This network determines the level of the E_0 parameter. One can see that the variation of this value depending on the plasticizer concentration has a completely different character when using DEHS and TBP.

We determined that $m = 0.33$ for E_0 in all samples swollen in DEHS. Also the same m value for E_0 was obtained for the PT-1 series swollen in the both plasticizers. But this value has a much higher level for other samples swollen in TBP (m =2.3). This fact suggests that DEHS does not practically destroy the network of strong physical bonds in segmented elastomers, the concentration variation in E_0 value corresponds to the geometric factor characteristic for swelling network of chemical bonds. The PT-1 series samples do not contain hard blocks. The variation of their parameters during swelling corresponds to the amorphous elastomers [8]. TBP, in contrast to DEHS, destroys partially the domains of hard blocks leading to the correspondent degradation of the spatial network in the investigated polymers during swelling. This effect is reflected by the higher level of the m value for E_0 parameters for segmented polyurethane urea elastomers.

Table 3. Parameters of equation (1) for investigated samples in the unswollen state

№	α	γ_s	b
PT-1	0.16	0.06	0.4
PT-2	0.16	0.06	0.4
PT-3	0.16	0.06	0.4
PT-4	0.16	0.06	0.4

4. Conclusion

It was found that the experimental dependence of stress versus strain for cross-linked segmented elastomers swollen in plasticizers can be successfully described using the equation previously tested for the non-swollen elastomers. Experimental confirmation of this fact was given for four series of segmented polyurethane ureas samples with different chemical structures and network density. The degree of the plasticizer structure influence on the level of variation in the relaxation parameters of the studied elastomers, and the deformation behavior of the swollen elastomers was found as well.

Acknowledgements

This work was financially supported by the Russian Foundation for Basic Research (Project 17-03–00119).

References

[1] Tereshatov V V; Senichev V Yu. *J. of Macromol. Sci. Part B: Physics* **53** 575-871
[2] Senichev V Yu, Strelnikov V N, Tereshatov V V, Makarova M A. *J. Macromol. Sci. Part B: Physics* **57** 196-209
[3] Tereshatov VV, Senichev V Yu. *Vysokomolek. Soed.* B **32** 848-51.
[4] Prisacariu C 2011 *Polyurethane elastomers from morphology to mechanical aspects* (Wien, New York: Springer-Verlag) p 255
[5] Tereshatov V V, Makarova M A, Senichev V Yu. *Int. Polym. Sci. Techn.* **34** 11–15
[6] Tereshatov V V; Makarova M A; Senichev V Yu; Slobodinyuk A I. *Coll. Polym. Sci.* **290** 641-51
[7] Bartenev G M, Lyalina N M. *Vysokomolek. Soed.* **12** 368-75
[8] Tereshatov V V, Senichev V Yu. *Vysokomolek. Soed.* A **37** 1888-95
[9] Bueche F. *J. Polym. Sci.* **24** 189-200

IOP Conf. Series: Materials Science and Engineering **665** (2019) 012012 doi:10.1088/1757-899X/665/1/012012

Magnetocaloric effect in thin magnetic films γ-Fe2O3

X O Urinov[1], X A Jumanov[1], Kh B Mirzokulov[1], A M Xidirov[1], M A Bolbekov[1], T A Jumaboyev[1] and J O Urinov[1]

[1]Samarkand branch of Tashkent University of information technologies named after Mukhammad al-Khwarizmi, 47A Sh.Mirzo Street, Samarkand, 140100, Uzbekistan

e-mail: liverpool_2592@mail.ru

Abstract. In this paper, we studied the thermal properties of thin films associated with significant difficulties caused by their small volume. Nevertheless, this problem provides undoubted interest, since calorimetric studies can provide useful information about the thermometric properties of films, phase transformations in them. The problem can be solved by using materials phase transition the metal-insulator in thermal contact with the sample as heat-sensitive elements of the films. The theory of the magnetocaloric effect in uniaxial and polycrystalline ferromagnets taking into account the rotation of magnetization is presented. The issues of creation of new refrigerating machines using the magnetocaloric effect are discussed. The magnetocaloric effect in magnetically ordered single crystals is studied taking into account the processes of magnetization rotation. An expression for the magnetocaloric effect in a ferromagnetic polycrystal is obtained, which well describes the experimental results.

Introduction

The study of thermomagnetic phenomena in magnets is one of the directions in the field of physics of magnetic phenomena. Under thermomagnetic phenomena is understood as a set of effects arising under the action of the magnetic field $\vec{H}$, when there is a heat flux of some density and the transformation of energy (both reversible and irreversible) magnetization into heat is realized[1]. The magnetocaloric effect (MCE) belongs to the class of reversible thermomagnetic phenomena.

The magnetocaloric effect is understood as the temperature of the magnet under the action of an external field. As a rule, in conditions of thermal magnetization increases the temperature, and the demagnetization - downwards. In principle, thermo magnetic energy devices phenomenon characteristic of all magnetic materials (para, ferro, antiferromagnetic materials), although the magnetocaloric effect is particularly large in ferromagnetic materials and paramagnetic materials[2].

Experimental study of thermal properties of thin films is associated with significant difficulties caused by their small volume. However, this problem provides great interest, as the calorimetric studies can provide useful information on the thermometric properties of the films, phase transformations in them, etc. the Problem can be solved by the use as thermosensitive elements of the films of materials with phase transition the metal - insulator (PTMI) being in thermal contact with the studied sample. Planar execution of the test film and the sensing element provides maximum heat transfer between the layers, and a strong change in the electrical conductivity of the films in the PTMI allows high sensitivity to record temperature changes. In this paper, we investigated the MCE in thin magnetic films . The PTMI in magnetically ordered crystals consists in temperature change at adiabatic change of an external magnetic field [3,4]. As will be shown below film geometry, the state of the surface and the nature of the magnetic anisotropy affect the appearance of MCE in thin magnetic films.

2. Magnetocaloric effect in thin magnetic films in the model of uniform rotation of magnetization

Consider MCE in thin magnetic films in a model of uniform rotation. The density of the free energy of the film, placed in an external magnetic field of intensity H, can be represented as:

$$f = f_0 + K\sin^2\theta - MH\cos(\psi - \theta) \tag{1}$$

where f_0 – the density of the free energy does not depend on magnetic state; K– constants of magnetic anisotropy of the film including anisotropy of the form and induced magnetic anisotropy if they are present in the sample; M- the magnetization of the film; and θ- accordingly, the angles between the selected direction and the directions of the magnetic field and magnetization vectors [5,6].

The value of MCE is determined by the slope of the isentrope

(2)

where S - entropy, $S = -{df}/{dT}$.

Using (1) and the condition of equilibrium of magnetization we obtain

$$\frac{df}{d\theta} = K\sin 2\theta - MH\sin(\psi - \theta) = 0 \tag{3}$$

we can show that

$$\frac{\partial S}{\partial H} = \frac{\partial K}{\partial T}\frac{\partial \theta}{\partial H}\sin 2\theta - \frac{\partial M}{\partial T}\frac{\partial \theta}{\partial H}H\sin(\psi - \theta) - \frac{\partial M}{\partial T}\cos(\psi - \theta) - \frac{\partial \theta}{\partial H}M\sin(\psi - \theta) \tag{4}$$

$$\frac{\partial S}{\partial T} = \frac{\partial S_0}{\partial T} + 2\frac{\partial \theta}{\partial T}\left[\frac{\partial K}{\partial T}\sin 2\theta - \frac{\partial M}{\partial T}H\sin(\psi - \theta)\right] + \left(\frac{\partial \theta}{\partial T}\right)^2 [2K\cos 2\theta + MH\cos(\psi - \theta)] \tag{5}$$

Considering the temperature change at MCE to be small, we can limit ourselves to the linear decomposition of magnetization and the constant of magnetic anisotropy in temperature [7]. Derivatives of θ angles by temperature and field are found from the equilibrium condition $df/d\theta = 0$.

$$\left(\frac{\Delta T}{\Delta H}\right)_S = \frac{T}{C_H}\frac{K\sin^2 2\theta}{MH[2K\cos 2\theta + MH \bullet \cos(\psi - \theta)]} \bullet D_3 \tag{6}$$

where $D_3 = \left(M\frac{\partial K}{\partial T} - K\frac{\partial M}{\partial T}\right) - \frac{T}{C_H}\frac{\partial M}{\partial T}\cos(\psi - \theta)$, C_H - heat capacity of the film in a magnetic field.

(7)

In the absence of magnetic anisotropy (K=0, ψ=0) expressions (6) go to the known for isotropic ferromagnet [8,9]. Depending on the ratio between the temperature behavior of magnetization and the magnetic anisotropy constant, the first term b (6) provides heating or cooling of the film. Note that in the absence of induced magnetic anisotropy (K=2πM^2) it is always negative.

$$2K\cos 2\theta + MH\cos(\psi - \theta) = \frac{d^2 f}{d\theta^2} \geq O \tag{8}$$

In this case, in particular C_H=C_0 for illstration in Figure 1. presents the angular dependence of the FE film with K=2πM^2, H=4πM.

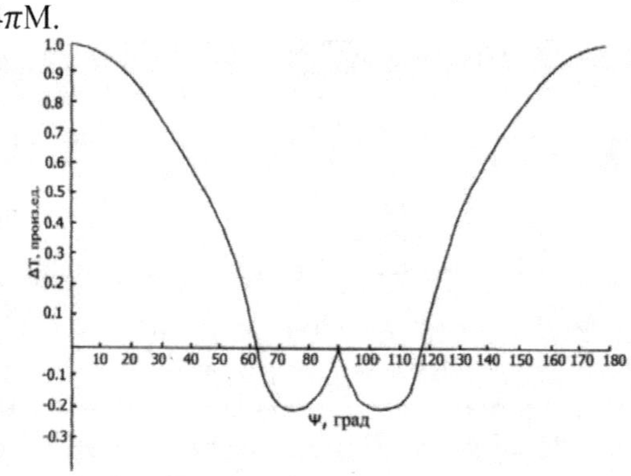

Figure 1. Anisotropy magnetocaloric effect in thin magnetic films in the model of uniform rotation of magnetization, K=2πM^2, H=4πM.

3. Sample preparation. Experimental technique

As objects of study were selected ferrimagnetic film of Nickel metal and ferrimagnetic insulator γ-Fe2O3. Thickness of samples was 1000 – 3000 A. Monocrystalline film γ-Fe2O3 received by the method of chemical transport reactions of substrates of magnesium oxide, the thickness of the substrates is 0.3 mm. as the temperature sensing element used the film of vanadium dioxide VO2. Vanadium dioxide at a temperature of 340 K and undergoes PTMI with a jump of conductance reaching four orders of magnitude [10]. Deposition of VO2 films was carried out by pyrolysis of vanadium acetylacetonate. To study FE in the films γ-Fe2O3 the VO2 deposition was carried out directly on the surface of the ferromagnetic; to eliminate the shunt effect in the case of metal films deposition of VO2 was performed on the opposite side of the substrate . The thickness of VO2 films was 1000 - 5000A.

The resulting layered systems were placed in a thermostatic chamber, inside which a flat furnace, thermocouple, inputs for electrical and power devices were mounted. The chamber was located in the gap of the electromagnet, the magnetic field strength of which varied within 2-20kE.

Electrical resistance of VO2 films was measured by two-probe method. Contacts to the samples were made using conductive glue based on polyacrylic resin. For the present work it was sufficient to have VO2 films with a jump in conductivity of about two orders and a temperature coefficient of resistance near PTMI – 3 kω/ deg.

It is established that the temperature change in the chamber when the magnetic field is turned on due to the possible magnetoresistive effect of the wire heater of the furnace is not manifested in the experiment. The magnetoresistive effect in VO2 films is negligible. This makes it possible to uniquely associate the temperature change in the layered system containing the magnetic film with the change of the magnetic field with the MCE in the latter.

The experiment was conducted as follows. At a fixed temperature and a given angle ψ, the magnetic field of a certain intensity was switched on and the change in the electrical resistance of the VO2 film was recorded. The temperature setting time was a few seconds, after that, due to nonadiabaticity , the temperature slowly returned to its original value.

4. Results and discussion

Using the described technique, we detected and investigated FEM in Nickel and γ-Fe2O3. Measurements were made on a series of films. In all cases, qualitatively unambiguous results are obtained. The peculiarity of the effect is its unexpectedly large value, characteristic, anisotropy and alternating signs in a certain area of angles and fields. For Figure 2 shows the angular dependence of MCE in the film γ-Fe2O3 thickness of 3000 A in a magnetic field of 5.3 kE. A characteristic feature of the effect in this case are two additional maxima ΔT, indicating the contribution of magnetic crystallographic anisotropy to the MCE, the first constant of which in γ-Fe2O3 is negative [5].

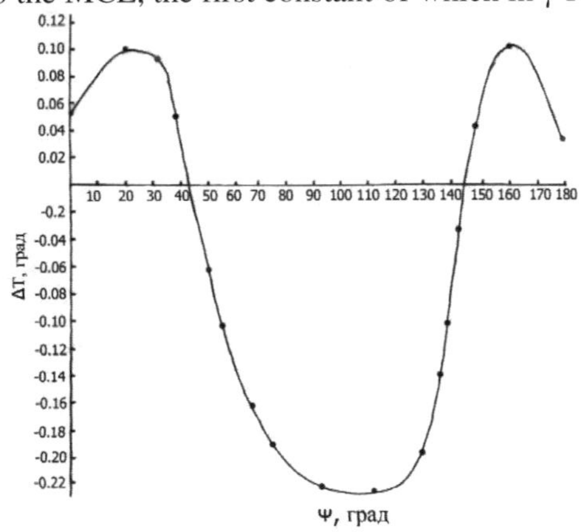

Figure 2. The angular dependence of the magnetocaloric effect in the film γ-Fe2O3 film thickness of 3000 A, the magnetic field of 5.3 kE

As in the case of film it is cooling with increasing tension, isentrope MCE ≈ - $2,4 \cdot 10^{-5}$ deg/E.

From the comparison of the calculated and experimental results it can be seen that the model of homogeneous rotation qualitatively explains the data. This refers to the anisotropy of the effect and its alternating sign.

5. Conclusion

In this work, we first conducted a pilot study of the finite element method in thin magnetic films. This was facilitated by the method proposed by the authors, based on heat exchange in a layered structure containing the investigated film and a film of a thermo sensitive material with PTMI. The measurements were conducted on films of Nickel and γ-Fe2O3, showed that a decisive role in the manifestation of MCE plays the character of magnetization of the films. The proposal of the authors with a significant contribution to the MCE of surface magnetic anisotropy does not contradict the results of the experiment.

References

[1] Belov K P and Nikitin S A 1989 *Magnetic properties of crystalline and amorphous media* (Novosibirsk: Nauka)
[2] Tishin A M and Spichkin Y I 2003 *IOP Publishing Ltd* pp 150-152
[3] Andreev V N and Klimov V A 2010 *Solid state physics* **52 (3)** pp 557-563
[4] Nikitin S A, Tskhadadze G A, Ovtchenkova I A, Zhukova D A and Ivanova T I 2010 *Proc. EASTMAG-2010* p 66
[5] Babkin E V and Urinov X O 1990 *Solid state physics* **32(4)** pp 689-694
[6] Babkin E V, Baranov G I and Urinov X O 1991 *Soviet technical physics letters* **17(5)** pp 10-12
[7] Urinov X O, Salaxitdinov A N and Mirzokulov Kh B 2018 Proc. New in magnetism and magnetic materials (XXIII international conference) pp 593-595
[8] Urinov X O Jumanov X A Salakhitdinov A N Nasimov X M Mirzokulov Kh B and Xidirov A M 2019 *Scientific journal of SamSU* **1** pp 90-93
[9] Oleynik A S 2002 *Technical physics* **72(8)** p 88

IOP Conf. Series: Materials Science and Engineering **665** (2019) 012013 doi:10.1088/1757-899X/665/1/012013

Martensitic transformations in VO2 films

X O Urinov[1], X A Jumanov[1], A K Amonov[1], A M Xidirov[1], J R Kilichov and J X Nurmurodov[1] T A Jumaboyev[1]

[1]Samarkand branch of Tashkent University of information technologies named after Mukhammad al-Khwarizmi, 47A Sh.Mirzo Street, Samarkand, 140100, Uzbekistan

e-mail: adkham1972@gmail.com

Abstract. Martensitic transformation is considered to be non-diffusive since the atoms in the transition are moved to a distance of the order of interatomic, and neighboring atoms in the initial phase remain adjacent and in the new one, only their position changes. In this paper, for the first time, an experimental study of the phase transition metal-dielectric (FPMD) in thin films of vanadium dioxide is accompanied by a structural transition of martensitic type, so this material is a convenient model object that allows you to clearly identify both the processes of defect formation and their influence on its physical properties in the vicinity of the phase transition point. The obtained results are naturally associated with the effect of adhesion and epitaxy on the phase transition, which can be reduced to the contribution to the energy balance of the component σ/d, where σ is the density of the surface energy; d is the film thickness. Physically, this may mean that since the growth of films occurs at a temperature above the FPMD temperature, the orientation of the crystal axes of the high-temperature tetragonal phase is stabilized by epitaxial forces from the substrate side, and this makes it difficult to reconstruct the crystal structure during the phase transition. Phase stabilization in thin layers due to surface effects is a common phenomenon in the physics of thin films.

Introduction

Martensitic transformation is a phase transition of the first kind, occurring by the origin and growth of a new phase, fully or partially coherent with the original. Metaphase boundaries are sources of internal stresses caused by the difference between the specific volumes of the initial and new phases. The products of martensitic transformation are polysynthetic phases, which are packages of plane-parallel layers – elastic domains [1]. Elastic domains can serve as a variety of crystalline phases and orientational variants of the same phase. Martensitic transformation is considered to be non-diffusive since the atoms in the transition are moved to a distance of the order of interatomic, and neighboring atoms in the initial phase remain adjacent and in the new one, only their position changes.

Martensitic transformations, under which, at the present time, I understand the whole class of phase transitions, got its name from the martensite – a very hard phase a kind of leaf shaped, formed, in the hardening by rapid cooling from the austenitic state of the alloy based on Fe-C. Under these conditions, cooling eutectoid diffusion and the release of carbon from solid solution becomes impossible. Under these conditions, the thermodynamic phase is more stable than austenite and can be obtained by the shear movement of atoms, in which a solid solution of carbon in iron with a face-centered cubic the lattice is converted into a solid solution of carbon in iron with a volume-centered lattice – martensite.

In body-centred cubic latticethe iron dissolved is not more than 0.02 %; in the martensite, the carbon concentration is the same as in austenite, i.e. it can reach 1.7 %. But at the same time the lattice, remaining volume-centered, cubic lattice turns into a tetragonal one with an ordered arrangement of carbon atoms.

The transformation of austenite into martensite is characterized by an increase in the specific volume of steel by 1 % (it is the more carbon in steel). Therefore, as already mentioned, the transformation is accompanied by a sharp increase in the elastic deformation of the lattice; the accumulation of elastic energy prevents the transformation and under certain conditions may not allow it to complete [2]. To

IOP Conf. Series: Materials Science and Engineering **665** (2019) 012013 doi:10.1088/1757-899X/665/1/012013

continue the transformation in this case, it is necessary to create an additional driving force by further lowering the temperature or applying a favorable sign of external deformation.

The effect of the subsequent reduction of the tetraganolicity of irradiated martensite as a result of short-term cooling in liquid nitrogen or rapid heating to room temperature was unexpected. Lowering tetragonality in this case happens abruptly and is a kind of phase transition of martensite. Such uncharacteristic non-irradiated martensite change at low temperature parameters of the lattice is not associated with since these transitions do not require carbon diffusion over long distances, they occur within the unit cell. The presence in the martensite lattice of point defects created by irradiation stimulates the transition of carbon atoms from one lattice position to another, energetically more favorable at these temperatures. The validity of this statement is evidenced by the results of the experiment, showing the chemical stability of irradiated martensite compared with non-irradiated (under the same conditions), since radiation defects prevent the diffusion of carbon atoms over long distances and thereby prevent the formation of carbides.

In this respect, thin films are a kind of analogue of irradiated martensite, since they are inherent in structural imperfections associated with structural imperfection of the substrate with a mismatch of the parameters of the film lattice and the substrate, etc., as a result of which the compound in the thin-film and massive States may have different properties under the same conditions. The above-mentioned defects can serve as places of localization of relaxing impurities. Good adhesion of films on substrates prevents the destruction of crystals and the growth of defects in the martensitic transition. Therefore, in addition to the mechanical strength of the film, they are stable in physical and chemical terms.

The phase transition metal-dielectric (PTMD) in thin films of vanadium dioxide is accompanied by a structural transition of martensitic type, so this material is a convenient model object, which allows to clearly identify both the processes of defect formation and their influence on its physical properties in the vicinity of the phase transition point[3,4,5].

2. Experimental technique and sample preparation

One of the most important characteristics of $V0_2$ is the temperature of PTMD T_k=340 K. As noted in the review, the all-round pressure leads to an increase in TC, cationic substitution can lead to both an increase and a decrease in it depending on the impurity, etc.

Here are the results of the study of the effect of the substrate and film thickness on the temperature of PTMD in V02 [6].

V02 films were obtained by pyrolysis of vanadium acetylacetonate, the low deposition temperature of which allowed toobtain films on various substrates. Mica, Sitall, corundum (Al2O3), lithium fluoride (LiF) and magnesium oxide (Mg0) Plates were used as a substrate. Upon receipt of the films were kept constant optimal substrate temperature 360 K and pressure of oxygen because of the synthesis of 50-60 mm of mercury the deposition Rate of films is 0.1 μm/MIN. The contacts were made by applying and then firing silver paste, and the distance between the contacts was 5 mm.

The result of this work was to establish the fact that the temperature of PTMD TC in VO2 films is always lower than for massive material, table 1. For rice. 1 the results of measurements of the temperature dependence of electrical resistance of films with a thickness of 0.47-0.51 microns on different substrates are shown. Here, for comparison, the temperature dependence of the resistivity of a massive crystal is given.

The temperature of the PTMDon the same substrates is monotonically decreasing with decreasing film thickness, So the temperature of the PTMD films on mica and magnesium oxide with a thickness of 0.02 and 0.01 μm, respectively, is 317±1 and 316±1 In Fig. 2 shows the thickness dependence of the temperature of PTMD films on a magnesium oxide substrate.

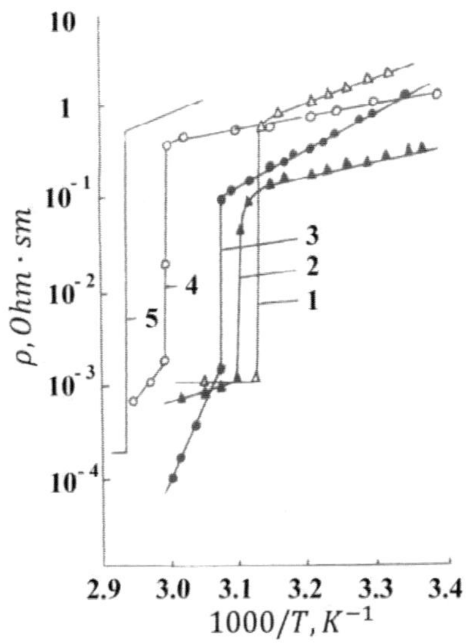

Figure 1. Temperature dependence of the resistivity of the films VO_2толщиной of 0.47-0.51 microns on various substrates: 1-MgO; 2-LiF; 3-Al_2O_3; 4 - sital; 5-data for the solid crystal [3].

The obtained results are naturally associated with the effect of adhesion and epitaxy on the phase transition, which can be reduced to the contribution to the energy balance of the component σ/d, where σ is the density of the surface energy; d is the film thickness. Physically, this may mean that since the growth of films occurs at a temperature above the PTMD temperature, the orientation of the crystal axes of the high-temperature tetragonal phase is stabilized by epitaxial forces from the substrate side, and this makes it difficult to reconstruct the crystal structure during the phase transition. Phase stabilization in thin layers due to surface effects is a common phenomenon in thin film physics [7,8].

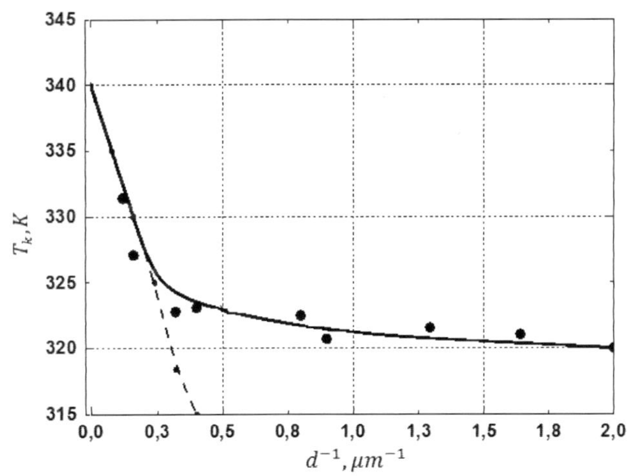

Figure 2. The temperature PTMD films VO2на substrate of magnesium oxide.

3. Results and discussion

Consider the phenomenological state of the film and the massive crystal at the phase transition [7]:

$$F_1(T_k) = F_2(T_k)\,,\quad F_1^0(T_k^0) = F_2^0(T_k^0)\,; \tag{1}$$

where F is the density of free energy; index "o" characterizes the properties of a massive crystal; indices "1" and "2" respectively refer to the metal and dielectric phases. The equations of state of the film and the massive crystal can be written in the form

$$U_1 + \frac{\sigma_1}{d} - T_k S_1 = U_2 + \frac{\sigma_2}{d} - T_k S_2 \tag{2}$$

$$U_1 - T_k^0 S_1 = U_2 - T_k^0 S_2 \tag{3}$$

where U – the internal energy density; S - entropy.

Obviously, the form of writing the phase equilibrium equation (I) imposes a limit on the lower limit of film thickness; this means that the contribution of surface energy to the total free energy should be small and should not change the thermodynamic properties of the film material. The limitation on the applicability of the model should be sought from the comparison of experimental results with the calculated data. From equations (I) and (3) it is easy to obtain an expression for the phase transition temperature of the film:

$$\mathbf{T_k = T_k^o - \Delta\sigma T_k^o / dQ} \tag{4}$$

where $\Delta\sigma = \sigma_2 - \sigma_1$ is the difference between the surface energy densities of both phases;

$$Q = \frac{(S_2 - S_1)}{T_k^0} \tag{5}$$

latent heat of phase transition.

As shown in Fig. 2 the experimental results are well described by the expression (3) starting with a film thickness of 3 μm. From the T_k (I/d) graphin particular, it is possible to estimate the difference between the surface energy densities of both phases. Using the value for the transition heat in VO_2 Q=1.02 kcal/mol [9] , we obtain $\delta\sigma = 2 \cdot 10^7$erg/cm^2 for the case of films on a magnesium oxide substrate.

Table 1. Temperature of FPMD VO2 films on various substrates

Substrate	$T_к$,K	Filmthickness, μm
MgO	320 ± 1	0,51
LiF	323 ± 1	0,50
Al_2O_3	325 ± 1	0,50
sitall	333 ± 1	0,47
Mica	333 ± 1	0,49
Massivecrystal	340 ± 1	–

Thus, the martensitic mechanism of transformation into steel plays an important role not only in the practice of metallurgy, but also in the fields of science and technology. In recent years, the metal - dielectric phase transition in semiconductor oxides of doped impurities of various kinds has been interpreted often involving the martensitic transition mechanism.

Conclusion

In this paper, an experimental study of FPMD in VO2 films of different thicknesses on different substrates was carried out for the first time. Studies of PTMD in VO_2 films of different thicknesses on different substrates. It is shown that the decrease in the transition temperature in the films is due to surface phenomena at the film-substrate interface.

References

[1] Roytburd A L 1977 *Martensitnie prevrasheniya i klassifikatsiya teorii fazovix perexodov* (Martensitic transformations and classification of phase transition theory) (Kiev: Naukova dumka)

IOP Conf. Series: Materials Science and Engineering **665** (2019) 012013 doi:10.1088/1757-899X/665/1/012013

[2] Grelik S S and Dashevskiy M Y 1973 Materialovedenie poluprovodnikov i metallovedenie (Semiconductor materials science and metallurgy) (Moscow: Metallurgiya)

[3] Babkin E V, Urinov Kh O, Chariyev A A and Dolgarev A A 1987 *The solid films* **150** pp 11-14

[4] Horlin T, Niklevaki T and Nyoren M 1973. *Mat. Res. Rull.* **7(2)** pp 179-189

[5] Urinov X O 2016 *Proc. Actual problems of molecular spectroscopy of condensed matter Solid state physics* pp 117-118

[6] Reyes J M, Segal S L and Sayar M A 1975 *Journal of solid state chemical* **17(5)** pp 298-302

[7] Bugayev A A, Zaxarchenya E P and Chudnovsky F A 1979 Metal - semiconductor phase transition and its application (Leningrad: Nauka)

[8] Urinov X O Jumanov X A Salakhitdinov A N Nasimov X M Mirzokulov Kh B and Xidirov A M 2019 *Scientific journal of SamSU* **1** pp 90-93

[9] Oleynik A S 2002 *Technical physics* **72(8)** p 88

Workstations Industry 4.0 for instrument engineering products

D A Zakoldaev[1], A V Shukalov[1], I O Zharinov[1], O O Zharinov[2]

[1] Faculty of Information Security and Computer Technologies, Saint Petersburg National Research University of Information Technologies, Mechanics and Optics, 49, Kronverksky Av., Saint Petersburg, 197101, Russia
[2] Department of Problem-Oriented Computing Complexes, Saint Petersburg State University of Aerospace Instrumentation, 67, Bolshaya Morskaia str., Saint Petersburg, 190000, Russia

e-mail: mpbva@mail.ru

Abstract. The actual task is to design an automatic work place to be applied in the Industry 4.0 to prepare electronically the item designing component technical documentation. An automatic work place is a part of Industry 4.0 digital factory instrument ware. Digital factory is a new type of project company functioning with innovation technologies. Here is an analysis of technical, program and algorithmic points of item projection in the existing companies of the Industry 3.0. There is a functional scheme of designer automatic work place in the Industry 3.0 company. We can see that new progressive innovative technologies in project companies require new approach to organize an automatic work place. The given functional scheme of designer automatic work place in the Industry 4.0 company requires implementation into project procedures cloud calculation technologies, industrial internet of things and other.

1. Introduction

Item designing components projecting is a very important direction how to implement new digital informative technologies [1, 2] which is a part of concept known as Industry 4.0. Industry 4.0 means to create [3, 4] in project companies an informative environment and technical means to complete the construction objectives using digital methods of modelling and virtualization.

The approaches of the Industry 4.0 concept is based [5, 6] on projecting instruments division into physical level components and virtual level components (cloud environment). Physical level components have origin in the projecting theory and practice of the Industry 3.0 and some new automatizing instruments in addition. Virtual level components [7, 8] are formed within the company cloud environment and include cloud services to automatize the designer project activity.

The most important designer project activity automatizing element is [9] the method of imitation modelling as a project procedure which is a good change for physical tests with model. Modelling tests [10] mean load application to the item digital model being conducted on the PCs of high productivity.

An instrument PC (Personal Computer) is the core of the designer automatic work place for item designing components and being integrated with program and hardware resources which designers apply in their project activity. The constitution of automatic work places for designers of the Industry 3.0 company and the Industry 4.0 company are quite different being based on the projecting technologies accepted into the company.

2. Automatic work places in the Industry 3.0 company

The main result of the designer project activity in the Industry 3.0 company is the technical documentation for the item designing component which include:

- construction documentation (CD);
- program documentation (PD);

- technological documentation (TD).

Item technical documentation is prepared electronically with special software systems of automatic projecting (CAD) installed in the instrument PC. CAD software is an application software which grants the designer a limited range of projection instruments and being functioned with the software control.

CAD application software let the designer complete the following project procedures in the Industry 3.0 company:

- engineering calculations (for example, the calculation of radio elements parameters of the electric circuits being created by the designer as electric scheme);
- construction modelling (for example, the behavior simulation of digital scheme components being watched by the designer in the instrument PC screen as a time chart);
- creation of technical documentation (for example, the preparation of part drawings under formal rules for this kind of document from branch and state standards);
- the compilation of program initial code (for example, the preparation of file components being downloaded into processor of the items controlled by the software and other).

Designer automatic work place (AWP) functional scheme for item designing components being used in the Industry 3.0 project activity is given in figure 1.

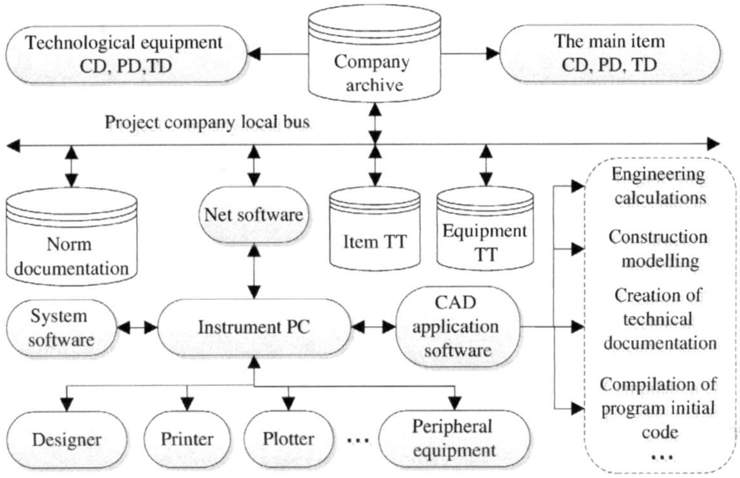

Figure 1. AWP functional scheme to project the item designing components in the Industry 3.0 company.

The completion of project procedures in the Industry 3.0 company requires the documents of processes and results of the designer project activity. The main way how to make documents of the project activity results is a set of peripheral printing equipment (printer, plotter and other) to create the technical documentation for the item designing component in paper. Documents in paper in the Industry 3.0 are used to create the exploitation documentation, additional notes of tests results for some stages of research and development activity, the sheets of concord for some construction documents and other.

All AWP for the designer project activity in the Industry 3.0 company are united into company local bus functioning with special net software installed in each instrument PC. Company net resources unite science and research and construction and technology departments with the company technical archive where the designer may access remotely. Company technical archive has in paper and electronically the following documents:

- construction, program and technological documents for the items being developed in the company to be manufactured later and delivered into exploitation (primary items);
- construction, program and technological documents for the items (technical accessories) which are special appliances which are necessary during the primary items manufacturing;

- norm documentation which is for accepted in this branch development rules and technical documents creation to control for designer the primary project procedures;
- technical tasks (TT) for primary items and technological accessories which have structural requirements of the ordering part for each type items and other.

3. Automatic work places in the Industry 4.0 company

Project activity result of the item designing component designer in the Industry 4.0 company is the electronic form of the item description. Electronic form of the item description is the item digital model with full geometry and properties of the item relevant for the exploitation. There are several forms of the item digital model:
- digital replica;
- digital shadow;
- digital twin,

to define the number of iterative project procedures done by the designer with electronic documents.

AWP core in the Industry 4.0 is an instrument PC connected to the communication environment of the project company functioning on the industrial Internet of Things (IoT) standards. The company communication environment is supported with the multi-agent access mode for designers to the company resources which are necessary for the project activity.

Designer automatic work place functional scheme for item designing components being used in the Industry 4.0 project activity is given in figure 2.

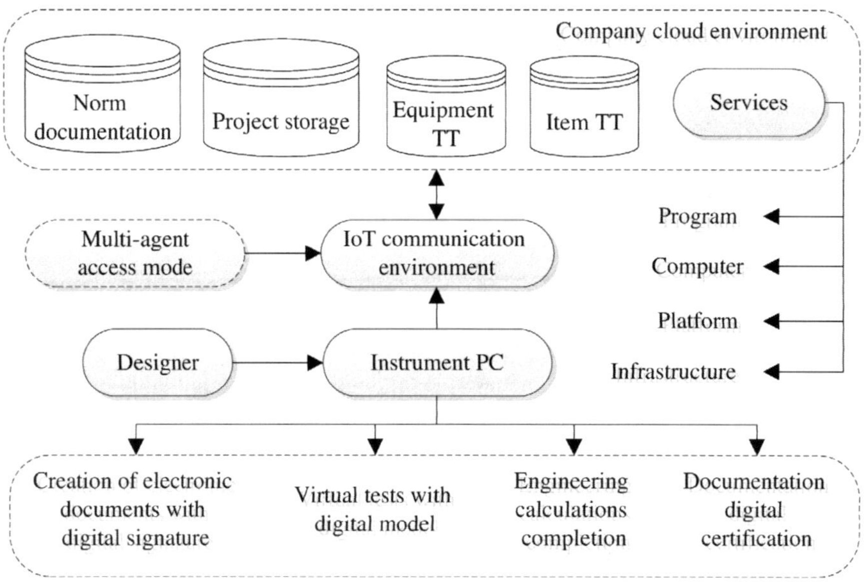

Figure 2. AWP functional scheme to project the item designing components in the Industry 4.0 company.

Company resources are placed in the cloud environment and have the following components:
- norm documentation (state and branch standards which are adapted for the company functioning in the digital economy);
- storage of projects (SQL (Structured Query Language)-like data base where in structural order CD, PD, TD of the primary items and their technological accessories are placed);
- technical task for the item development done in a format of text file, for example xml);
- cloud services as resources of program, computer, platform, infrastructure and other types.

An AWP let the Industry 4.0 company designers complete the following types of project procedures:

- to create electronic documents with digital signature of document executor and technical documentation quality norm controller;
- to conduct virtual tests with digital model to find out the project errors in the first stages of the item model creation;
- automatized or automatic completion of engineering calculations including the heat and electromagnetic fields calculations of the items being developed;
- technical documentation digital certification with the evaluation of correspondence to the requirements of the item digital model to the standards of the digital economy and other.

4. Conclusion

Project procedures of the Industry 3.0 and the Industry 4.0 companies today is based on the same type of technical means and different software for project activity. Projection automatizing technical means provide the designer a set of instrument functioning under software control component to prepare the documentation for the item. Projection automatizing program means have instrument to research different options of project solutions to choose the optimal one according to the technical task criteria for item realization option.

Item designing components projection instruments accepted in the Industry 4.0 is a result of digital informative technologies development defined the furthest technical progress in different fields of industry. The most important are the instruments of projection realized as cloud services which are capable to automatize the designer activity on the base of cyber and physical systems application.

A cyber and physical system is a combination of hardware and software components which together is capable to make higher the level of automatizing of project activity and production activity in the Industry 4.0 industrial companies. The development of mathematical system and software for cyber and physical system will help to improve the project procedure of the item designing components creation and to increase the quality of the item digital models. Item digital models is a form of technical documentation being developed in the Industry 4.0 digital factory and transferred to the Industry 4.0 smart factory to be manufactured.

References

[1] Qian Zh, Yu Y, Fan G 2015 *Applied mathematics & information sciences* **9** 1981-92
[2] Centobelli P, Cerchione R, Murino T and Gallo M 2016 *International journal of simulation modelling* **15** 223-235
[3] Stef I D, Draghici G, Draghici A 2013 *Procedia technology* **9** 451-462
[4] Shariatzadeh N, Lundholm Th, Lindberg L and Sivard G 2016 *Procedia CIRP* **50** 512-517
[5] Gjeldum N, Mladineo M, Veza I 2016 *Procedia CIRP* **54** 158-163
[6] Ruttimann B G, Stockli M T 2016 *Journal of service science and management* **9** 458-500
[7] Sarkar A, Sormaz D 2018 *Procedia manufacturing* **17** 435-443
[8] Gurjanov A V, Zakoldaev D A, Shukalov A V and Zharinov I O 2018 *IOP Conference Series: Materials Science and Engineering* **327** 022110
[9] Zakoldaev D A, Shukalov A V, Zharinov I O and Zharinov O O 2018 *Journal of Physics: Conference Series* **1015** 052033
[10] Jeong H-Y, Jeong Y-S, Park J H 2014 *Sustainability* **6** 8510-21

Workstations Industry 4.0 for instrument manufacturing

D A Zakoldaev[1], A G Korobeynikov[1], A V Shukalov[1], I O Zharinov[1]

[1] Faculty of Information Security and Computer Technologies, Saint Petersburg National Research University of Information Technologies, Mechanics and Optics, 49, Kronverksky Av., Saint Petersburg, 197101, Russia

e-mail: mpbva@mail.ru

Abstract. The actual task is to project the structure of operator automatic work place in the Industry 4.0 smart factory. The Industry 4.0 smart factory is a new type of production company. Smart factories are oriented to manufacture item designing components in automatic mode. The key technologies of humanless and paperless production are cloud technologies, industrial Internet of things, sensors technologies and others. A principle is given how to construct automatizing work place in existing production companies of the Industry 3.0. There is a functional scheme of the Industry 3.0 production company automatized work place. It is clear that implementation of new informative technologies influence significantly the structure of an operator automatized work place in production. The proposed functional scheme of the automatized work place is for the Industry 4.0 smart factory.

1. Introduction

The organization of item designing components production in the Industry 4.0 company is based [1, 2] on implementation into technological processes the cyber and physical systems and technologies of humanless production. Cyber and physical systems [3, 4] is a new type of production machines with two primary components:

- physical component which is the hardware itself to produce a limited number of technological operations inside the work chamber;
- virtual (cyber and physical) component placed into company cloud environment to support informatively the item manufacturing technological processes.

The technologies of humanless production applied in the Industry 4.0 [5, 6] require the implementation of the following components into technological processes of item manufacturing [7, 8]:

- electronic documents of construction (CD), program (PD), and technological (TD) documentation based on the net informative technologies;
- production machines which are cyber and physical systems of industrial purpose which are capable to complete production tasks automatically;
- self-organization algorithms of cyber and physical systems placed into industrial company cloud environment to ensure the completion of technological processes of item designing component manufacturing without humans and other.

The division of company production resources into physical and virtual components is a stage [9] of the Industry 3.0 company development where the base for item manufacturing are physical components [10, 11]. To project innovative productions of the Industry 4.0 must be done in special automatized work places of operators oriented to use the progressive industrial technologies.

2. Automatized work places to produce item designing components in the Industry 3.0 company

Central component of the automatized work place (AWP) to produce the item designing components in the Industry 3.0 company is the technical support which include:

- instrument PC (Personal Computer) with installed system, net and application software to check, make settings and adjustments of the items being manufactured automatically (with human participation);
- peripheral printing equipment which make documents of the completed technological operations being done on the item by the manufacturer;
- technological models of unity to interact with instrument PC interface through cable connection and check remote control for the item being manufactured.

The item designing components manufacturer AWP functional scheme of the Industry 3.0 company is given in figure 1.

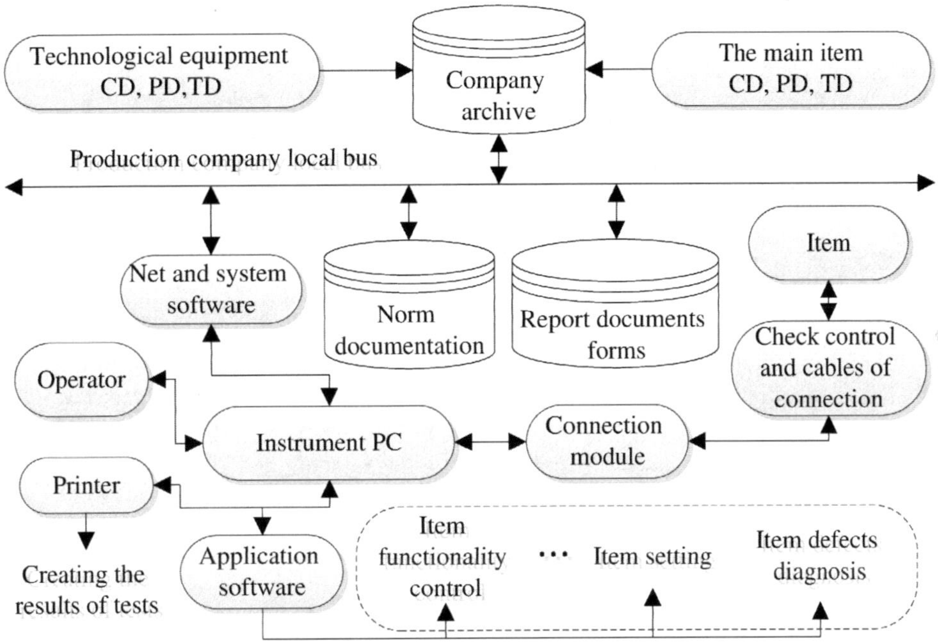

Figure 1. Item designing components manufacturing AWP functional scheme of the Industry 3.0 company.

Manufacturer (operator) with application software tells the instrument PC the volume of item check which depends on the number of program algorithms. Item check modes could be of software or hardware means. Machine means to program item check modes are remote control organs which is a part of AWP technical means. System, net and application components are a part of AWP software means.

The application software helps the operator in the Industry 3.0 production to do the following:
- correctness control of the item being manufactured which is done by internal or external hard ware and software means;
- setting of the item being manufactured being done by the operator automatically to make sure that item functional and accuracy properties as they should be and relevant for the exploitation;
- item faults diagnosis manufactured in production, in exploitation to find out the reasons of faults (defects) after which the item has broken and other.

Item manufacturing technological operations completion results in the Industry 3.0 companies are documented by the operator with the forms of report documents (test protocols, fault analysis acts, entering control acts, fault acts and other.). The report documents forms are according to the company standards or to branch (state) standards. The company operator may find all necessary reporting document forms into company electronic archive.

Production company operator access to the electronica company resources is done through the net software interface installed in instrument PC. All instrument PCs which are part of AWP united in the Industry 3.0 company into local bus where also electronic archive is connected. Along with report documents forms and norm documents (state standards) company electronic archive also has CD, PD, TD of the item being manufactured and accessories (technical accessories) which are necessary for manufacturing. CD, PD, TD give the operator informative support of his activity processes in the company to produce the item designing component of high quality.

3. Automatized work places to produce item designing components in the Industry 4.0 company

Automatized work place of the item designing component manufacturer central component in the Industry 4.0 company is a cyber and physical system. The cyber and physical system (CPS) hardware include:

- work chamber to complete technological process (for example, 3D-printing of parts from powder material);
- a set of CPS condition detectors and detectors of technological manufacturing processes completion quality control into the work chamber;
- a controller which with special software may form digital commands which influence the gear to reach the necessary modes of the technological operations completion;
- a actor (electric and mechanical unit) which does the commands from controller to maintain the necessary instrument movements trajectories inside the work chamber (for example, a nozzle of 3D-printer).

Each CPS may complete in production a limited number of technological operations. Unification of CPSs in their acceptability for production process into clusters may create special sections (workshops) with necessary functional capabilities. CPS informative interactions united into technological sections is done in communication environment of industrial company.

Communication environment of the Industry 4.0 is a channel of wireless high speed connection functioning in multi-agent access mode for the specialist and CPSs. CPS hardware provide the information exchange in communication environment with program and software means of controller. Specialist (tester) of the Industry 4.0 production company may access to the communication environment resources through the interaction interface available in mobile devices (tablet, phone, communicator, computer and other.). Communication environment of the Industry 4.0 industrial company support the standards of information exchange with protocols of Internet of Things (IoT).

Company informative resources placed into cloud environment include the following components:

- norm documentation (a set of state and branch standards oriented to production processes being done into digital economy conditions);
- CD, PD, TD of the item being manufactured represented as item digital models;
- algorithm libraries of CPS self-organization which helps CPS to interact with multi-agent technologies;
- methods and forms of test protocols being conducted on the item being manufactured to prove the item quality and other.

The item designing components manufacturing AWP functional scheme of the Industry 4.0 company is given in figure 2.

The Industry 4.0 AWP helps the operator in production to do the following technological operations:

- production processes completion with humanless and paperless technologies automatically;
- conduct physical tests with manufactured item in automatized (automatic) mode;
- remote creation of electronic protocols about manufactured item physical tests results;
- manufactured item digital certification with standards of digital economy.

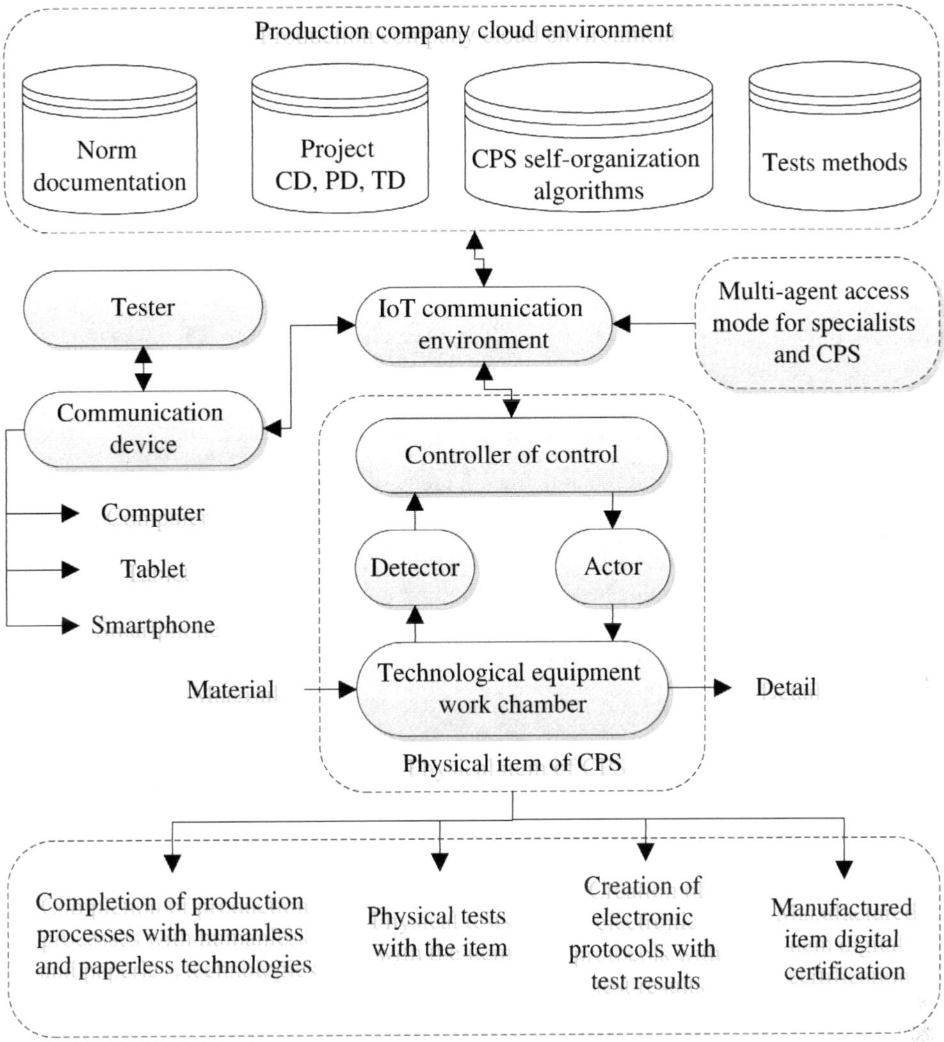

Figure 2. Item designing components manufacturing AWP functional scheme of the Industry 4.0 company.

4. Conclusion

To project a production division of the Industry 4.0 with highly-productive CPSs and innovative technologies is a relevant direction of modern industry development. The main options how to create a smart factory could be realized in the following ways:

- by modernization of production capabilities and existing science and research institutes and item designing bureaus of the Industry 3.0;
- by construction of new plants equipped with cyber and physical systems and which have qualified personnel trained to work with CPSs.

If we are to modernize the existing capabilities of the Industry 3.0 companies we must stage by stage renew the technical equipment and company software with the components functioning in the digital space. Also they need professional re-training of personnel in special educative organizations accredited to prepare the personnel with new industrial occupations.

Educative programs to prepare the specialists of the Industry 4.0 must have material and technical support which is adequate to the progressive industrial technologies (cloud calculations, Internet of

IOP Conf. Series: Materials Science and Engineering **665** (2019) 012015 doi:10.1088/1757-899X/665/1/012015

Things, security of cyber and physical systems, big amount of data analysis, sensors and other.) which are being applied today in item and machine designing.

If we are to construct new plants (smart factories) we must assess the company economic criteria and also how rentable a high-tech production with high added value is and which is the most suitable option to develop the item designing industry within the Industry 4.0.

References

[1] Jabeur N, Sahli N, Zeadally Sh 2015 *Mobile information systems* **2015** 908315

[2] Hwang G, Lee J, Park J and Chang T-W 2017 *International journal of production research* **55** 2590-02

[3] Muller R, Vette M, Geenen A 2017 *Procedia Manufacturing* **11** 13-21

[4] Lee J, Noh S D, Kim H-J and Kang Y-Sh 2018 *Sensors* **18** 1428-44

[5] Nitschke C S, Williams S P 2018 *Procedia computer science* **138** 580-586

[6] Gurjanov A V, Zakoldaev D A, Shukalov A V and Zharinov I O 2018 *IOP Conference Series: Materials Science and Engineering* **327** 022110

[7] Zakoldaev D A, Shukalov A V, Zharinov I O and Zharinov O O 2018 *Journal of Physics: Conference Series* **1015** 052033

[8] Yu J, Liu J, Xiang L and Zhou J 2017 *Procedia IUTAM* **22** 107-114

[9] Nyanga L, Van der Merwe A F, Matope S and Dewa M T 2015 *Procedia CIRP* **28** 167-172

[10] Wu W, Li W, Law D and Na W 2015 *Procedia engineering* **118** 1266-73

[11] Bogatyrev V A, Parshutina S A, Poptcova N A and Bogatyrev A V 2016 *Communications in Computer and Information Science* **678** 337-348

The life cycle of technical documentation in the smart factory of Industry 4.0

D A Zakoldaev[1], A V Gurjanov[2], A V Shukalov[1], I O Zharinov[1]

[1] Faculty of Information Security and Computer Technologies, Saint Petersburg National Research University of Information Technologies, Mechanics and Optics, 49, Kronverksky Av., Saint Petersburg, 197101, Russia
[2] Director, Stock Company «Experimental Design Bureau «Electroavtomatika» named after P A Yefimov, 40, Marshala Govorova St., Saint Petersburg, 198095, Russia

E-mail: mpbva@mail.ru

Abstract. The task is to synthesize all stages of the item designing component technical documentation life cycle in the companies of the Industry 4.0. The companies of the Industry 4.0 is a new type of project and production companies which activity is done with paperless documentation. The traditional designing procedures of construction, program and technological documentation are studied which are available today in the companies of the Industry 3.0. There is a scheme of technological documentation life cycle in the Industry 3.0 companies. It is clear that for the Industry 4.0 companies they need new forms of item designing component technical documentation which could be digital replica, digital shadow and digital twin of the item which is done in different stages of item projection. Also it is clear that new forms of electronic documents require new instrument means for project activity. There is a scheme of item designing component technical documentation life cycle in the Industry 4.0 companies.

1. Introduction

Improvement of methods and means of projection and production for the item designing components led [1, 2] to appearance of new information technologies which is the base of digital and smart factories of the Industry 4.0. Digital and smart factories are of deep automatizing of project and production procedures of the item creation which are available now thanks to the technologies of electronic documents [3].

Company electronic documents requires a special informative and telecommunication environment [4, 5] the key components of which are machines and software means of data exchange. Electronic documents are to support most of business processes [6] in the company including the projection of technical documentation.

Technical documentation projection technologies of today for the item designing components are based [7, 8] on the usage of special systems of computer aided design which helps the designer to prepare all types of documents which are parts of construction, program and technological documentation. Those technologies are instead now of paper documents which require all schemes, drawings and text documentation to be prepared without any CAD (Computer Aided Design) mean.

For the Industry 4.0 [9, 10] companies the technologies of electronic documents is only the beginning which will be developed by implementation into project and production activity the new forms of item properties description as a digital product. Those new forms of description require mathematical theories and methods which show the components and material properties as equations and non-equations.

Digital product properties mathematical forms of description may lead to the development of imitation modelling and virtualization which helps to change in project and production companies of the Industry 4.0 some business processes including the item manufacturing and its natural tests.

The implementation of digital product properties mathematical forms of description into project and production activity alongside with electronic documents require to re-establish some stage of the item life cycle from an Industry 3.0 company. They need to re-asses the construction, program and technological documentation and in the future it will help to create item designing component digital replica, digital shadow and digital twin.

2. Technical documentation life cycle in the Industry 3.0 company

Item designing component technical documentation in the Industry 3.0 company is done electronically and classified in three types of documents:
- construction documentation (CD);
- program documentation (PD);
- technological documentation (TD).

The development of technical documentation is done in the stage of the item structure projection according to the requirements of the technical task (TT).

Construction documentation is for description of mass and dimensions characteristics of the item which are done with standards with rules how to create those documents.

Construction documentation is a set of electronic documents which includes:
- schemes;
- drawings;
- text documentation

which a designer creates with computer aided design systems of projection and technical means which is a part of automatic work place in a project company.

Program documentation describes algorithms of item functioning in exploitation and it has some special industry standards. Program documentation is a set of files and programs which include:
- algorithms;
- text documentation;
- machine code of the files being executed

which is done by the designers of the items controlled by program.

Technological documentation describes production processes of item manufacturing and it has some special industrial standards. Technological documentation is done after its construction and program counterparts. The technological documentation includes:
- schemes;
- drawings;
- text documentation

which is prepared for the processes of item manufacturing and for technological equipment of the production process. The content of technological documentation describes which actions the Industry 3.0 company worker must do to create the item itself and its assembly units.

The technical documentation quality for the Industry 3.0 item designing components is evaluated by:
- a norm controller who checks each type of documents of CD, PD, TD which is a set of rules and norms in standards;
- a designer who conducts a set of tests with an item sample manufactured after its CD, PD, TD in the company production division.

Results of visual assessment of technical documentation each type and results of natural tests which show some discrepancies in quality of item functioning from the one of technical task could be a good base for some modifications of CD, PD, TD. The iterative process of technical documentation quality assessment in the Industry 3.0 company must be repeated until the following results are achieved:
- rules how to make all documents of CD, PD, TD are completed according to the state and branch standards;
- item properties which are results of test with its quality and quantity evaluation according to the requirements of this item development technical task.

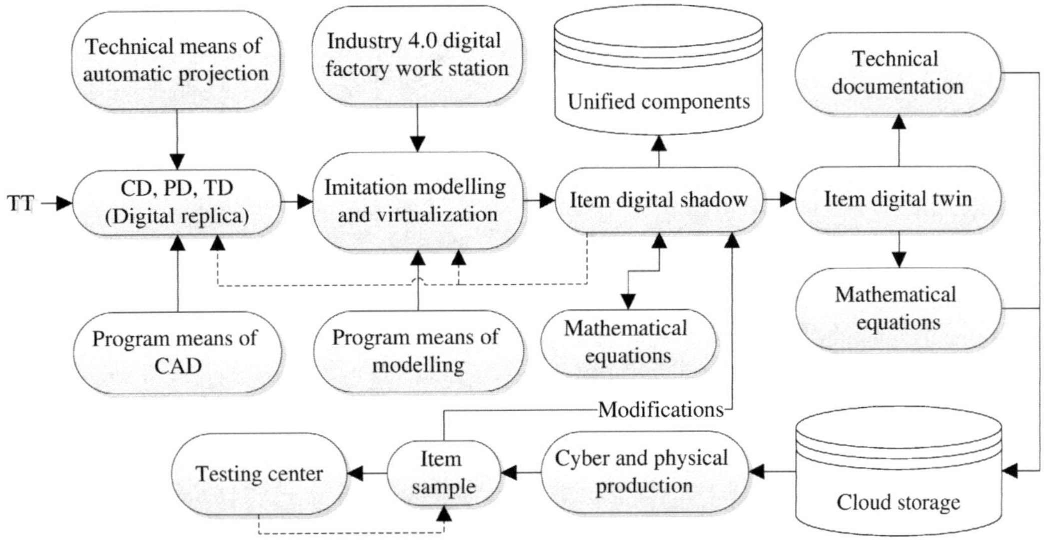

Figure 1. Technical documentation life cycle scheme in the industrial companies of the Industry 3.0

The scheme of technical documentation life cycle being done in the Industry 3.0 company is given in figure 1. Having analyzed the figure 1 we can say that item construction and program documentation being developed in the Industry 3.0 company in parallel according to the requirements of technical task. Technological documentation is being developed after its construction and program counterparts.

The correction of technical documentation after the each type of documents by a norm controller analysis results which are done separately for CD, PD, TD. The changes to CD, PD, TD after its sample tests are done at the same time.

3. Technical documentation life cycle in the Industry 4.0 company

Item designing component technical documentation made in the Industry 4.0 company is a digital product of the company projection activity which is represented in one of the following forms:

- item digital replica which is technical documentation like CD, PD, TD done electronically after project procedures which are quite like the procedures of the Industry 3.0 company;
- item digital shadow which is its CD, PD, TD with some mathematical equations to describe its physical and chemical properties of materials and its components which is the part of the item;
- item digital twin which is the ideal form of item description done first as a detail accepted after item digital shadow.

Forms to describe items as digital replica, digital shadow and digital twin is a specific thing of the Industry 4.0 and manifest the implementation of new knowledge in different fields of applicable and fundamental sciences into project activity of a digital factory.

To prepare the item digital replica electronically they use computer aided design systems and technical means to automatize the project activity which are like the components available now in the Industry 3.0 company.

Item digital shadow is made after modelling of the item 3D-model and its assembly units into a virtual environment which is a software of work station of the Industry 4.0 digital factory. Modelling tests of digital shadow is done according to the requirements of the external factors influences and its technical task for the item development.

Positive results of the digital shadow modelling tests is the base to create the item sample in cyber and physical production. The item sample is tested with natural tests in the test center of a digital factory to approve the properties and technical characteristics of the item being good for the requirements of the item development technical task. Results of modelling tests of the item digital shadow and results of physical tests of the item sample is a reason to modify technical documents of the item digital shadow.

Item digital shadow is an item description form from which the documentation will be transferred to the Industry 4.0 smart factory for its serial production. Serial items being exploited have some loads which are higher than the ones from the technical task so that why the further improvement of the item documentation is required which will be supported through the whole life cycle of the item.

Notes after the item exploitation (a bunch of items) is a reason to change the item digital shadow and after some iterations the digital twin will be created which includes:

- technical documentation (CD, PD, TD);
- mathematical equations which are good for real exploitation conditions to describe masses and dimensions of the item and its physical and chemical properties.

Digital replica, digital shadow and digital twin is a result of separate project activity in the Industry 4.0 company and they are stored in the company cloud storage. Technical documentation life cycle scheme in the companies of the Industry 4.0 is given in figure 2.

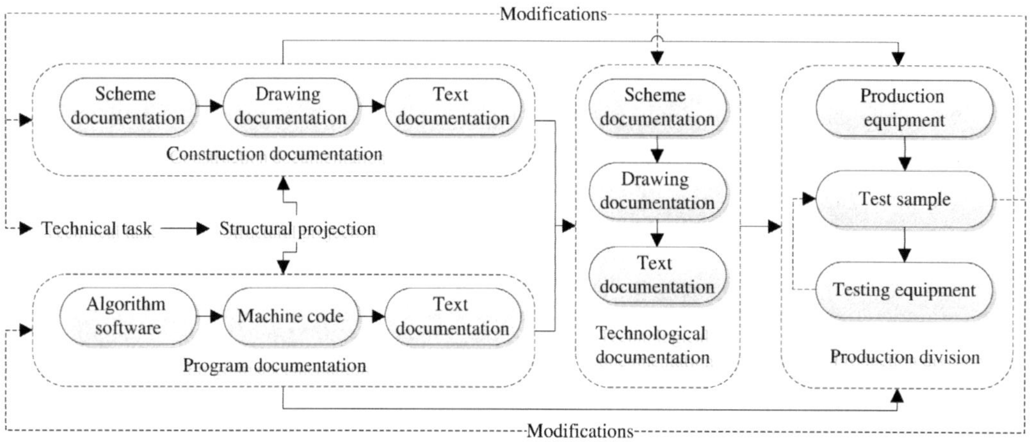

Figure 2. Technical documentation life cycle schemein the industrial companies of the Industry 4.0

4. Conclusion

To describe item properties as mathematical equations is a perspective direction for the item technical documentation life cycle stages development being done in the Industry 4.0 company paperlessly and humanlessly. Main components of those mathematical equations for the item designing components are differential and integral non-equations to define physical laws of optics, radio signals electronics, mechanics, thermodynamics and other.

The components of mathematical equations (non-equations) is a part of the item 3D-model according to its construction, program and technological documentations to describe masses and dimensions of the item and its physical and chemical properties and also optical, cinematics and other properties of the assembly units.

To describe an item as its mathematical model will help to create the libraries of assembly units bases (assembly units of wide application) which helps to reduce the time of manufacturing and to increase the quality of the items being developed. Mathematical equation in this case is a universal mean which can be implemented into the program environment of modelling and virtualization to support the test procedures of the item digital models.

References

[1] Zhuang C, Liu J, Xiong H 2018 *The International Journal of Advanced Manufacturing Technology* **96** 1149-63

IOP Conf. Series: Materials Science and Engineering **665** (2019) 012016 doi:10.1088/1757-899X/665/1/012016

[2] Lee E A 2015 *Sensors* **15**(3) 4837-69

[3] Alam K M, Saddik A E 2017 *IEEE Access* **5** 2050-62

[4] Gurjanov A V, Zakoldaev D A, Shukalov A V and Zharinov I O 2018 *Journal of Physics: Conference Series* **1059** 012010

[5] Haag S, Anderl R 2018 *Manufacturing Letters* **15** 64-66

[6] Jabeur N, Sahli N, Zeadally Sh 2015 *Mobile information systems* **2015** 908315

[7] Kritzinger W, Karner M, Traar G, Henjes J and Sihn W 2018 *IFAC PapersOnLine* **51** 1016-22

[8] Gurjanov A V, Zakoldaev D A, Shukalov A V and Zharinov I O 2018 *Journal of Physics: Conference Series* **1015** 052035

[9] Bogatyrev V A, Parshutina S A, Poptcova N A and Bogatyrev A V 2016 *Communications in computer and information science* **678** 337-348

[10] Hwang G, Lee J, Park J and Chang T-W 2017 *International Journal of Production Research* **55**(9) 2590-2602